DESIGN *of* COST-EFFICIENT INTERCONNECT PROCESSING UNITS

Spidergon STNoC

SYSTEM-ON-CHIP DESIGN AND TECHNOLOGIES

Series Editor: Farhad Mafie

Low-Power NoC for High-Performance SoC Design
Hoi-Jun Yoo, Kangmin Lee, and Jun Kyoung Kim

*Design of Cost-Efficient Network-on-Chip Architectures:
Spidergon STNoC*
Miltos D. Grammatikakis, Marcello Coppola, Riccardo Locatelli,
Giuseppe Maruccia, and Lorenzo Pieralisi

DESIGN *of* COST-EFFICIENT INTERCONNECT PROCESSING UNITS

Spidergon STNoC

Marcello Coppola
Miltos D. Grammatikakis
Riccardo Locatelli
Giuseppe Maruccia
Lorenzo Pieralisi

CRC Press
Taylor & Francis Group
Boca Raton London New York

CRC Press is an imprint of the
Taylor & Francis Group, an **informa** business

CRC Press
Taylor & Francis Group
6000 Broken Sound Parkway NW, Suite 300
Boca Raton, FL 33487-2742

© 2009 by Taylor & Francis Group, LLC
CRC Press is an imprint of Taylor & Francis Group, an Informa business

No claim to original U.S. Government works

ISBN 13: 978-1-4200-4471-3 (hbk)

Library of Congress Cataloging-in-Publication Data

Design of cost-efficient interconnect processing units : Spidergon STNoC /
   authors, Miltos D. Grammatikakis ... [et al.].
      p. cm. -- (System-on-chip design and technologies ; 2)
   "A CRC title."
   Includes bibliographical references and index.
   ISBN 978-1-4200-4471-3 (hardback : alk. paper)
      1. Networks on a chip. 2. ST Microelectronics. 3. Microprocessors. I.
Grammatikakis, Miltos D. II. Title. III. Series.

TK5105.546.D47 2009
004.1--dc22   2008026558

Visit the Taylor & Francis Web site at
http://www.taylorandfrancis.com

and the CRC Press Web site at
http://www.crcpress.com

Disclaimer

The views discussed within this book and contained in the ancillary CD are completely personal and do not reflect any opinions or policies of our affiliated institutions.

Moreover, concerning the on-chip communication network (OCCN) framework provided in the ancillary book CD, in no event shall any of the contributors be liable for any direct, indirect, incidental, consequential, exemplary, or special damages (including, but not limited to procurement of substitute goods or services; loss of use, data, or profits; or business interruption) resulting in any way from the use of the software framework. By adopting this specification, the user assumes full responsibility for its use.

Supplementary Resources Disclaimer

Additional resources were previously made available for this title on CD. However, as CD has become a less accessible format, all resources have been moved to a more convenient online download option.

You can find these resources available here: www.routledge.com/9781420044713

Please note: Where this title mentions the associated disc, please use the downloadable resources instead.

Dedication

We wish to express our sincere indebtedness by dedicating this book to our families for their constant support and encouragement throughout this long and tedious effort.

Contents

List of Figures

Foreword

A network-on-chip or NoC has become a key component in most complex digital chips. As the number of components per chip continues to double every two years all but the smallest chips today are *multicore* or *manycore* chips; that is, they consist of a number of separate *cores*, processors or fixed-function accelerators, along with memory arrays and input/output units. Mainstream general purpose processors are currently available as chip multiprocessors (CMPs) with four cores and tens of cores are expected by the end of the decade. Embedded systems-on-a-chip (SoCs) already have 10s of cores and will have 100s of cores within a decade.

Historically buses or crosspoint switches were used to connect multiple cores on a chip. Above four cores, however, the limited bandwidth of a bus becomes a major limitation, and above 8 or 16 cores, crosspoint switches become prohibitively expensive and unwieldy. Only a NoC provides the scalability needed for emerging multicore chips.

The NoC is a critical component of these emerging multicore and manycore CMPs and SoCs. They limit the performance realized by these chips and account for a substantial fraction of the power and area of these components. A recent NSF workshop identified the architecture and design of NoCs as a critical problem that must be solved to enable future CMPs and SoCs.

There is a rich literature on the design of system-level interconnection networks (c.f. Dally & Towles) and the design of these networks is well understood. While much of this system-level technology can be directly applied to NoCs, on-chip networks present a number of unique challenges that require solutions distinct from the tried-and-true system-level techniques.

Perhaps the biggest difference between system-level networks and NoCs is their cost structure. For system level networks, the bulk of the cost is in the channels. To first approximation the optimal system-level network is one that delivers the required performance with a minimum number of channels (and in particular a minimum number of long, and hence costly, channels). In contrast, the on-chip channels used to realize NoCs are constructed from inexpensive on-chip wires. The cost of a NoC is dominated not by channels, but rather by switches and buffers. Hence, the optimal NoC is one that minimizes switch and buffer area, often at the expense of more channels. This difference in cost structure motivates the use of very different topologies, routing algorithms, and flow-control methods in a NoC than would be used in a system-level network.

In addition to a completely different cost structure, NoCs also must deal with different traffic loads and use cases than seen by system-level networks. Many SoCs, for example, must deal with isochronous communication and provide QoS guarantees for certain types of traffic.

The unique cost and use constraints of NoCs pose many interesting research challenges. Hence, the architecture and design of NoCs has become an active research area with significant programs underway at Stanford, Princeton, Cambridge, and elsewhere. There is now an annual research conference devoted entirely to NoCs.

This book by Coppola and his colleagues is an excellent treatment of a timely topic. The authors start with an encyclopedic treatment of background topics ranging from CMPs to the basics of interconnection networks. They go on to describe the design of the Spidergon and the tools used to configure it in detail.

The Spidergon NoC is the first *industrial-strength* NoC to be described in detail. The book takes a system-level approach that leaves the reader with an understanding of not just how the Spidergon works, but also why the designers chose to make it work that way. Because the Spidergon is not just a NoC, but rather a family of NoCs and a methodology for configuring a particular instance of the family, the treatment gives the reader key insights into the issues faced in designing NoCs to meet a wide range of requirements.

This book is a must read for anyone involved in the design of SoCs, CMPs, or NoCs. It is both a snapshot of the current state of the NoC field and a unique look at the design of a state-of-the-art NoC. The reader will find themselves constantly referring back to the text to see how the Spidergon has dealt with issues ranging from routing to network interface design.

William J. Dally
Chairman of Computer Science
Stanford University
Author of: *Principles and Practices of Interconnection Networks*

Preface

The Network on Chip (NoC) is a simple packet-switched on-chip micronetwork foreseen as the natural evolution of traditional bus-based solutions, such as AMBA AHB and AXI, IBM CoreConnect, and ST Microelectronics STBus. Although Multicore SoC and NoC can apply techniques from embedded systems, parallel and distributed computing, and computer networks, new principles, constraints and processes are necessary for optimizing within the specific on-chip domain wire density and gate cost, reliability, performance, and power consumption. In addition, NoC has to be designed as a flexible integration platform by exploiting IP block- and system-level reuse at all development levels, since it is the only way to achieve current time-to-market requirements.

This research-oriented book (monograph and CD) is the result of years of collaborative work among all authors. More specifically, in six monograph chapters, we examine in a comprehensive, consistent and pragmatic way design aspects of Multicore SoC, and especially NoC design, unraveling a wealth of important theoretical and practical topics, such as bus versus point-to-point NoC complexity in deep submicron technology, NoC topology selection, component design (router, link, and interface), and innovative system-level methodology for efficient NoC modeling and design space exploration. Two special features of this book are that the monograph discusses architectural choice and implementation issues of ST Microelectronics' industrial NoC program (called Spidergon STNoC), while the ancillary CD introduces the only open-source SystemC-based on-chip communication network framework, providing practical training opportunities on NoC modeling.

Chapter 1 focuses on the growing trend for Multicore architectures, including algorithms and applications, providing implications for desirable interconnect features. In the context of convergence towards heterogeneous multicore architectures, the Interconnect Processing Unit (IPU) is introduced as a configurable and extensible communication component implementing system services and core communication. We introduce evolution from SoC to Multicore architecture, providing architectural concepts of current on-chip multiprocessor designs, such as PicoArray, Ambric AM series, Connec CA1024, Tilera, XMOS and Stanford's Smart Memories project. Multicore SoC software complexity is an important aspect of these novel embedded architectures, and we discuss fundamental concepts considering both user-space, i.e. application and middleware layers, and kernel space, i.e. real time operating system (RTOS), system libraries, hardware abstraction layer, device drivers and hardware layers. We relate software performance to Multicore programming and concurrency, and program correctness to coherency and consistency aspects. We also examine other sources of increased complexity for Multicore SoC architectures related to technological implications of submicron silicon technologies, including deep submicron effects, power consumption, clock synchronization, and supply power. Finally, we describe key concepts that drive efficient Multicore design, such as design for manufacturing, IP reuse, hierarchical design, and design regularity.

Chapter 2 examines wire-centric design, reporting an interesting chronological evolution of on-chip communication architectures as background to finally expose motivations towards the transition from on-chip buses to NoC, focusing on realistic industrial requirements. We outline the current state-of-the-art and features of several commercial on-chip buses, such as AMBA AHB and AXI, IBM CoreConnect, Sonics Silicon Backplane, and ST Microelectronics STBus. Then, we briefly examine design characteristics of existing regular point-to-point or customizable NoC frameworks, such as LIP6 SPIN, MIT Raw, University of Bologna's and Stanford's Xpipes, Philips Æthereal NoC, Silistix Chainworks, Arteris NoC, and ST Microelectronics V-STNoC.

Chapter 3 is devoted to definitions, theoretical performance, and cost metrics of NoC topologies. As a guideline for comparative analysis, we first discuss requirements and present theoretical metrics for on-chip network topologies. We examine the family of multistage interconnection networks (MINs), including blocking, wide- and strict-sense non-blocking. For constant degree, point-to-point architectures, we examine common 2-dimensional mesh and torus topologies. Then, we consider interesting circulants of degree 2, 3 or 4, and other constant degree graphs, such as cube-connected cycles, De Bruijn, and fat tree. Within this context, we finally introduce the degree-3 Spidergon topology by ST Microelectronics at the basis of Spidergon STNoC technology. Spidergon is a low diameter, symmetric circulant graph. This chapter concludes by reporting a comparative analysis based on the defined theoretical metrics. The Spidergon STNoC topology is studied against other common topologies, such as ring and 2-dimensional mesh, considering practical requirements for the on-chip NoC context, such as limited network size and fine grain network extendibility. These results show that higher degree topologies, such as 2-dimensional mesh, have higher cost, while being competitive in terms of network diameter and average distance to Spidergon topology only for square networks with a relatively large number of nodes. Hence, Spidergon is a cost-performance tradeoff NoC solution for the embedded SoC context, filling the gap between a simple and low performance ring and a highly connected mesh.

Chapter 4 elaborates on the architecture of the ST Microelectronics' Spidergon STNoC, an industrial NoC program first announced in December 2005. The Spidergon STNoC, a first attempt at designing an IPU (Interconnect Processing Unit), provides an innovative network-oriented platform into new demanding Multicore SoC applications. As an aggressive and innovative architecture philosophy, Spidergon STNoC leverages the regularity of Spidergon topology while featuring a pragmatic topological customizability. Following a simple, but clean and well-defined communication layering, the low-cost Spidergon STNoC architecture is based on three components: a wormhole router, a physical communication link, and a network interface providing uniform NoC access from any computing engine, memory, I/O, or application-specific IP core. First, we examine general aspects of the Spidergon STNoC architecture, such as the family of supported topologies, the adopted switching technique, communication protocol layering, packet structure, and finally, routing algorithms and deadlock avoidance. All different concepts are first discussed from a theoretical point of view, illustrating all possible alternatives. Then, specific choices are deeply motivated and explained. At the end of this chapter, we discuss fundamental design concepts and innovative features for Spidergon STNoC components: the packet router, the network interface and the physical link.

Chapter 5 examines electronic system-level design (ESL) as a crucial design methodology for complex Multicore SoCs with short time-to-market. We consider general ESL design methodology and tools, and we discuss basic concepts, such as IP reuse, multiple levels of abstraction, separation of functional specification from implementation, orthogonalization between communication and computation, communication layering, top-down, bottom-up and platform-based refinement, system- and transaction-level modeling. We then briefly discuss system-level NoC design space exploration tools for evaluating architectural alternatives, such as topological properties, performance, and energy efficiency. NoC tools often generate SystemC or HDL code (through high level synthesis) that can be used to estimate area and power consumption or provide adapters for simple integration of standardized IP blocks. Finally, we focus on the On-Chip Communication Network (OCCN) framework, the only open-source (GNU GPL license) system-level NoC modeling, simulation and design space exploration environment currently available. We conclude by describing future OCCN extensions.

Finally, Chapter 6 provides conclusions and discusses future trends in NoC design.

Although literature related to NoC is vast and fast-growing, the authors, describing their pragmatic research in the framework of the Spidergon STNoC program, have made a sincere effort to combine in this monograph a uniform, comprehensive, technically accurate overview of the current state-of-the-art and future trends in NoC design. Past and present literature is revisited from the industrial perspective suggesting an innovative, but practical, interpretation of the NoC concept. In addition, a large body of related references is provided to enable interested readers to pursue research in different areas of endeavour.

In addition to the monograph, practical training on efficient system-level modeling and design space exploration of complex NoC systems can be based on the supplementary On-Chip Communication Network (OCCN) software package developed and implemented by the authors which is available with the book CD. OCCN currently runs under Linux and Solaris operating systems and includes a SystemC-based library, technical documentation, and worked examples with point-to-point channel, bus and NoC models, for the normal user and channel designer. A generic NoC router model has been built using OCCN facilities in a few months.

Marcello Coppola	Miltos D. Grammatikakis	Riccardo Locatelli
Grenoble, France	Heraklion, Greece	Grenoble, France
Giuseppe Maruccia		Lorenzo Pieralisi
Grenoble, France		Grenoble, France

Acknowledgements

This book is the result of several years of hard work and creative thinking at ST Microelectronics that involved the authors in productive collaboration with many parties.

First and foremost we would like to thank our AST NoC colleagues who are currently working with us on the Spidergon STNoC project, contributing to the success and continuous evolution of this technology. Many thanks to Valerio Catalano, Michael Soulie, Philippe Teninge, Khaldoun Hassan, Esa Petri, Nico LInsalata, Michele Casula, Francesco Vitullo, and Nicola Concer. We would also like to express our sincere thanks to Sabina Fanfoni.

Moreover, the second author would like to express his gratitude towards colleagues at the Computer Science group at TEI-Crete, ISD S.A. and ST Microelectronics for maintaining a friendly and stimulating work environment.

During these years we have invested in creating a network of excellent universities, strongly based on human relationship and professional expertise. We warmly thank Prof. Luca Fanucci, Prof. Sergio Saponara, Prof. Luciano Bononi, and Prof. Luca Carloni. The authors want also to acknowledge Prof. Renaud Pacalet, Prof. Frederic Petrot, Prof. Gianluca Palermo, Dr. Davide Pandini, Prof. Cristina Silvano, and Prof. Jose Nunez-Yanez.

We would like to express our sincere gratitude to Prof. William J. Dally of the University of California at Berkeley who gave us the honor of reviewing this book, offering his tremendous experience in the field to promote our work and concepts. In addition, we would like to thank all our editor staff, in particular Nora Konopka, Marsha Pronin, and Allison Shatkin, who assisted us greatly and in a very courageous way during preparation of the final manuscript.

We would also like to extend our deep appreciation to our initial proposal reviewers, Prof. Massimo Conti, Dr. Grant Martin, and Prof. Cecilia Metra, for their exemplary review and helpful comments, provided in a short time interval.

This work has been partially supported by the MEDEA+ within the EU-REKA framework, the IST in the Sixth Framework Programme and the French Ministry of Economy, Industry, and Employment.

Biographies

Marcello Coppola is an experienced R&D Director at the Advanced Systems Technology (AST) Grenoble Lab of ST Microelectronics. Within this position he is leading the research strategy, and he is responsible for the development of the Spidergon STNoC program. He is joint inventor on 18 international patents. He studied computer science at Pisa University. In 1992, he received his Masters degree and started working with the Transputer architecture group of INMOS, Bristol (UK). For two and a half years he worked on a research program concerning the architecture of the C104 router. His research interests include several aspects of design technologies for System on Chip, with particular emphasis on network-on-chip, communication-centric architecture, Multicore, programming models, and ESL design. His publication record covers the field of simulation, IP modeling, SoC architecture, and on-chip communication networks. He has written chapters for different books. He was one of the members of the OSCI language working group. He contributed to SystemC2.0 language definition and OSCI standardization. He has chaired international conferences on SoC design and helped to organize several others. He has served as a Programme Committee member of several conferences, such as DATE, FDL, CODES+ISSS, MEDEA+, and DAC.

Miltos D. Grammatikakis received his MSc (1985) and PhD (1991) in Computer Science from the University of Oklahoma. He was postdoctoral fellow at the parallel processing laboratory LiP of ENS-Lyon (1991-1992), researcher at ICS, FORTH (1992-1995), assistant professor (C1) at the University of Hildesheim (1995-1998) and Oldenburg (1998-1999), senior software engineer and consultant at INTRACOM S.A. (1999-2000) and ISD S.A. (2000-), visiting professor the University of Crete (2002-2005) and associate professor at TEI (2003-). He has collaborated externally with ST Microelectronics for almost ten years, and participated in many European ESPRIT, IST, TEN TELECOM, TMR, EURESCOM, MEDEA+ and national R&D projects dealing with parallel architectures, EDA, high-level power estimation, distributed systems for telecom, and satellite networks. He has contributed to the ST Microelectronics proprietary IPSIM environment for system-level SoC modeling, the open source on-chip communication network framework (http://occn.sourceforge.net) for network-on-chip design space exploration, and the EDA Roadmap on IP Reuse. He has been selected as an expert reviwer for EU R&D projects. He has published more than 50 technical articles in edited books, journals and international conference proceedings, and is a co-author of *Parallel Systems: Communications and Interconnects* published by CRC. This book received the Editor's Choice Award in the IEEE Network column "New Books and Multimedia" in 2001. Dr. Grammatikakis has been a member of IEEE, ACM and SIAM. He has been cited in Marquis Who's Who in Engineering and IBC biographies.

Riccardo Locatelli received his Masters degree (summa cum laude) in Electronic Engineering in February 2000 and his PhD degree in Information Engineering from the University of Pisa in 2004. In 1999 he spent a research internship at the Microelectronics Section of the European Space Agency in the Netherlands, and in 2003 a visiting research at the Advanced Systems Technology (AST) Grenoble Lab of ST Microelectronics. At Pisa University he worked on the definition and prototyping of video architectures with emphasis on low power techniques and system communication. As a digital design engineer at CPR-TEAM, a microelectronic design house in Pisa, he worked on advanced signal processing schemes for VDSL applications. Since 2004, he has been at ST Microelectronics in Grenoble and is responsible for the architecture and design of the Spidergon STNoC IPU. His main research interests include all different aspects of network-on-chip from architecture to design, from tooling to simulation platform and performance analysis. He is a member of the technical program committee of NoC Symposium and DATE, and a reviewer of IEEE TCAD journal and several international conferences. He has published over 20 papers in international journals and conference proceedings and he has filed 10 international patents.

Giuseppe Maruccia received his Master degree in Electronic Engineering from the University of Cagliari in 2000 and immediately joined the Advanced System Technology (AST) Catania Lab of ST Microelectronics. Until the end of 2003 his research activity focused on SoC modeling, in particular development and verification of two SystemC-based modeling and simulation frameworks: IPSiM and OCCN (http://occn.sourceforge.net). Since 2004 he has moved to AST Grenoble Lab of ST Microelectronics, France, where he actively collaborates on the Spidergon STNoC design. In particular, until 2006 he was responsible for the network interface architecture specification. Currently, he is mostly concerned with the Spidergon STNoC validation and prototyping activity.

Lorenzo Pieralisi received his Master degree in Microelectronics (summa cum laude) in 2002 from the University of Ancona in Italy. He collaborated for a year with the Microelectronics Lab (DEIT) in Ancona, where he developed ad-hoc software based on SystemC for system-level power estimation. Since 2003, he has worked as a software engineer for the Advanced Systems Technology (AST) Grenoble Lab of ST Microelectronics, where he focuses on network-on-chip modeling and simulator development. He is currently heading network-on-chip modeling activities at AST Grenoble Research Lab, including distributed simulation and transaction-level modeling. He is a contributor and maintainer of OCCN, an open-source SystemC-based framework for network on-chip modeling and design space exploration, available at http://occn.sourceforge.net.

Chapter 1

Towards Multicores: Technology and Software Complexity

1.1 Multicore Architecture, Algorithms and Applications

Advanced semiconductor technologies have launched the second industrial revolution after the invention and deployment of steam engine and integrated assembly line a century ago. These technologies are based on the transistor invented in 1947 by Bell Lab's Bardeen, Stockeley, and Brattain, the integrated circuit (IC) built in 1958 by Texas Instrument's Jack Kilby, and the first all silicon chip built by Fairchild Camera's Robert Noyce in 1961.

While in the 1960s and 1970s, state-of-the-art semiconductor technology was only used by high-budget industries operating in the infrastructure arena, such as mainframe computers, telephone exchanges, and military systems. According to the Computer History Museum [77], in 1974, the first ever System-on-Chip (SoC) appeared in the Microma watch when Peter Stoll integrated the LCD-driver transistor and timer function in a single Intel 5810 CMOS chip. In the 1980s, memories, microprocessors and Application Specific Integrated Circuits (ASICs) based on CMOS technology scaling enabled a new electronics age. Leading-edge semiconductor technology penetrated emerging corporate applications, such as personal desktop computers (PCs), private telephone exchange systems (PBX) and fax machines, while the large consumer sector remained in the beginning "cheap and low-tech". During this period, the second generation of desktop computers, called home computers, was particularly popular with systems, such as Apple II, Atari 400/800, Commodore VIC20 and 64, and Sinclair ZX80, ZX81, and Spectrum.

In the 1990s many ASIC vendors started to address SoC opportunities by developing competitive products, such as hand-held games and instruments, digital satellite (DVB), speech processing, data communications, and PC peripherals. These products were based on a single embedded processor and hardware accelerator Intellectual Property (IP) blocks implementing specific application functions, such as H264, MPEG-4 encode/decode, and AAC stereo encode/decode, for improved cost-performance ratio.

This situation has changed dramatically during the past ten years. Continuous advances in deep submicron technology have led to multimillion transistor

IC designs that are incorporated into complex, faster, reliable and low-cost PCs and embedded systems for consumer products with a broad and growing diversity of application domains, such as entertainment, automotive, cellular phone, and set-top-box. For example, there are now approximately 1 billion PCs which represent less than 2% of all processors; remaining processors are used in embedded SoC devices.

This continuous progress in different directions always makes future prediction very hard. For almost half a century, SoC technology is strongly driven by Moore's empirical law [127] which predicts that the number of transistors placed on a chip approximately doubles every two years. For SoC design, this law forecasts an exponentially improved power-performance-price ratio. It also predicts for how long a product, e.g. a processor, memory or consumer electronic device can stay profitable in the world market, i.e. an implicit model of economic and social change. Thus, this law is the driving force for maintaining a healthy SoC market by integrating more application functions, while steadily reducing cost and time-to-market. For instance, the cost of a single transistor integrated on a memory chip has been reduced about one million times in the last 30 years.

Although precise market evolution is unknown, SoC is expected to provide new opportunities for large differentiation and product innovation through a continuous evolution in the market value chain [224]. In order to grow revenue and profits, key segments of the consumer electronics market value chain are expected to evolve over time through innovations in several core technologies, such as processor, memory, interconnect, operating systems, security, power management, graphics, and multimedia.

Convergence of system functionality and user experience is an evolutionary step which drives new terminals in the middle between PC and mobile phone (also known as feature-rich devices) to appear in the market. Thus, electronic consumer products sold previously as separate devices are currently being integrated together into a single electronic device, expanding opportunities through an ever-widening array of products. These devices include portable media players, personal navigation, digital radio receivers, portable games, car navigation systems and other consumer devices, capable of processing high quality media contents, such as decoding and encoding video (e.g. MPEG2, MPG4, H264), audio (e.g. AAC, MP3,WMA) and image (e.g. JPEG, gif). New projects involving photo-realistic video games include advanced multimedia data mining, artificial intelligence, instant high-definition video communications, and real-time speech recognition capabilities are other examples of this trend. Since many services are often provided together, e.g. Samsung's hybrid set-top box connects to a home PC or home media server exchanging multimedia content with a variety of consumer electronic devices, the above convergence imperative for SoC design leads to new silicon complexity and technology-related scalability issues, especially for power and memory bandwidth.

Convergence creates an opportunity for new markets in which innovation not only occurs in new types of products, but also in new kinds of business models, since traditional customers try to escalate their value chain. For example, Apple iPhone is a revolutionary product which includes a widescreen iPod, a built-in WIFI, a mobile phone, an internet communication device, and a digital camera associated to an innovative user interface. In regard to the business model, Apple not only sells iPhone hardware at a tidy profit, but also maintains complete control of its product line, collecting a percentage from telecom operators based on end-user utilization rates.

IBS market analysis indicates that R&D expenditures of IC vendors for SoC product development (or enhancements) are increasing as a percentage of revenues during the transition from the 0.13μm down to 22nm chip technology generation [165]. In particular, product development costs reach 30 percent of revenues for 22nm technology. Moreover, traditional SoC design flow typically requires two years between freezing a product specification and starting volume production, device manufacturers face increased pressure to introduce new models (usually software upgrades) every year, or even more frequently, adjusting to different and rapidly changing market and product scenarios. Thus, in order to reduce overall development cost, design risk, and non-recurring engineering (NRE) costs, more and more companies are outsourcing processor, on-chip interconnect and subsystem designs to third party specialists. Moreover, the differentiation factor of a final product is often linked to the quality of features delivered to the final user, e.g. the impressive touch screen user interface of the iPhone. Since the time evolution of such features is smaller than the current SoC design time, it is mandatory to build a SoC on top of a flexible product. In this context, products in the PC industry and embedded consumer market experience fundamental architectural changes while entering the Multicore era. The remainder of this Section examines trends in this evolution, already experienced in current PC products and application-specific embedded systems.

Recently, most major PC industries are shifting to multiple cores on a single chip to improve processor performance. Increasing the processor clock rate which has been a popular technique for obtaining speed up for over 30 years has run out of steam, due to excessive power consumption, heat dissipation, and electro migration-related reliability issues; a 1% clock speed up results in a 3% power increase [46]. In addition, memory bandwidth required by a processor is increasing due to higher clock frequency, while offered DRAM bandwidth is not growing with the same pace, causing large memory latencies. In order to alleviate this problem, more effective use of the memory hierarchy from processor to DRAM is introduced, including complex cache architectures, such as L3 cache.

Furthermore, in recent years, Instruction Level Parallelism (ILP) has also been exploited for the simultaneous execution of several instructions, possibly out of order. However, ILP relies heavily on complex hardware structures to achieve more work per processor cycle, such as speculative branch prediction, multiple instruction issue, dynamic scheduling, non-blocking caches, and deep pipelines. Moreover, increasing the clock-rate, more and more instructions could be executed, but, in modern applications, locating instructions that can be executed in parallel, e.g. by a compiler, is a very hard task. In fact, all the intrinsic ILP is nearly fully utilized by today's processors; therefore, there is no more possibility to increase performance as it has been achieved in the past.

For these reasons, as feature size decreases with new submicron technology, multicore processors are becoming more popular than large, power-hungry processors. Several **homogeneous Multicore processors** based on identical cores (processors) currently exist, such as Sun's 8-processor Niagara [344] and a newly announced version called Rock with 16 cores and several dual and quad cores: IBM's dual-core PowerPC 6, AMD's new quad-core Opteron server and workstation CPU [384], code-named Barcelona, and finally Intel's Tanglewood [163] and Cloverdale processors code-named Wolfdale (dual-core) and Yorkfield (quad-core). All these homogeneous Multicore processors (also known as chip-level multiprocessor, CMP) replicate several identical cores on the same die, interconnecting them, possibly through a hierarchy of cache and local memory, and accessing a (virtual) shared memory though one or more memory controllers. This traditional approach is application-independent, i.e. the same architecture can execute different types of applications. For example, AMD64 quad-core processor includes four cores, each with separate L1 and L2 caches and a common L3 cache accessing memory and I/O over an on-chip communication infrastructure [12], while Intel quad-core processor includes four cores, each with separate L1 cache, a shared, dynamically-allocated L2 cache across, while memory and I/O access over an advanced memory controller.

Although today inside a multicore there are usually up to two or four identical complex processor cores running at a lower clock rate to alleviate thermal issues, Multicore processors will soon contain eight or more complex processor cores. Success of these designs definitely depends on workloads that can leverage the additional processor cores. For example, it has been reported that for large-scale, fine grain thread-level parallel applications and multiprogramming workloads, multiprocessors can perform 50 to 100% better than a wide superscalar micro-architecture [258]. Independent of application domain, Multicores programming is no longer sequential, i.e. a fundamental revolution occurs towards mainstream concurrent software [342]. For this reason, new languages must exploit inherent parallelism, e.g. the Multicore initiative [250]. This revolution may actually be more important than the hardware one, since at least for certain applications, we can improve performance and reduce power by adding more cores.

SoCs for the embedded consumer market are customized to consumer applications. They are based on multiple processors on the same die, creating a new type of architecture that we refer to as **Multicore SoC**. Typical examples are the TI OMAP platform [384] and Philips MSVD [249]. These systems are characterized by heterogeneous cores coupled with hardware accelerators running at lower speed, and a specialized application-specific set of instructions. They offer the right level of performance, power dissipation, design flexibility, and cost compared to high performance solutions based on homogeneous Multicore processors.

Multicore SoC uses a control plane with mainly general-purpose host processors, such as ARM9, ARM11, ARM Cortex, MIPS, and StrongArm, and a data plane with application-specific data engines or streaming processors tuned to a specific class of data processing algorithms, e.g. MPEG4 encoding, decoding and transcoding digital audio. These processors are connected through an on-chip interconnect to multiple integrated low end peripherals, hardware acceleration blocks, registers, embedded storage components, and external DRAM memory controllers. This architectural approach enables application-specific design dedicated to intensive tasks, achieving orders of magnitude better performance and reduced power consumption over traditional general-purpose solutions. In fact, general-purpose processor cores execute all tasks at higher frequency, using much more memory, thus resulting in an inefficient solution. Current Multicore SoC architecture is different from Multicore processors. In the former case, applications and usage are known at design time, while in the latter case, these cannot be estimated in the PC market.

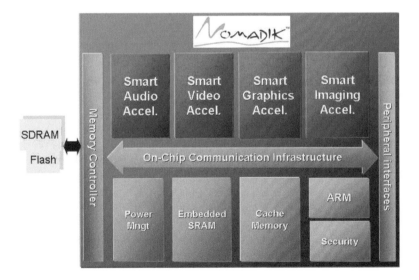

FIGURE 1.1: Nomadik: An example of a complex Multicore SoC

The block diagram of the ST Nomadik application processor for a 3G cell phone handset shown in Figure 1.1 illustrates this. The top-level blocks illustrate tailored programmable cores, with dedicated Direct Memory Access (DMA) services and hardware accelerators executing multiple non-interfering data intensive functions, such as audio, video, image or graphics processing. Other tasks, e.g. user interface functions, are performed in the general-purpose host processor, in this case an ARM. In addition to the global on-chip interconnect which represents a key system component, local buses (not shown) allow subsystems, such as video pipes to communicate to each other.

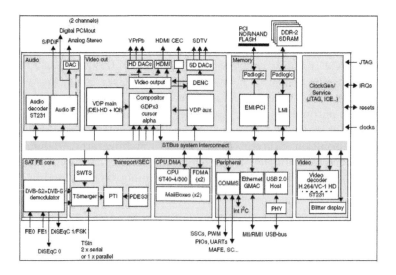

FIGURE 1.2:   Block diagram of HDTV satellite set-top-box decoder SoC

Another example shown in Figure 1.2 is a highly integrated HDTV satellite set-top-box decoder for VC-1, H264 and MPEG2 which include several processors for video and audio decoding, a host processor (an ST Microelectronics ST40) to run the application under different operating systems (Linux, Windows, and ST proprietary OS21), DVB-S2 and DVB-S front-end integrated demodulators, HD and SD simultaneous display, 2 video, 3 graphic planes and background color with high quality filtering, and many peripherals that handle memory and connectivity functions. Notice that all subsystems are connected through STBus, an on-chip bus developed in ST Microelectronics.

Between Multicore homogeneous processors and Multicore SoCs, we can identify yet another category of products called **Multicore heterogeneous processors**. Similar to Multicore processors, these heterogeneous architectures decouple the host controlling core from a set of specialized processing cores and are designed to be relatively open in terms of supported applications (targeting at least a class of applications). Well-known examples include the Intel network processor, and the IBM, Sony and Toshiba Cell broadband engine that addresses mainly the video game market [158]. The Cell is a well-known general-purpose pipelined Graphics Processing Unit (GPU) which can perform real-time computing, as well as graphics processing. In a general sense it can be positioned in the gap between general purpose CPU(s) and specialized high-performance processors, such as traditional graphic processors. It is based on a single 64-bit PowerPC processor surrounded by eight identical SPE co-processors. Similar to the central PowerPC processor, each SPE (synergistic processing element) is a dual-issue machine, but unlike the PPE (parallel processing element) the two execution pipelines are not symmetrical. In other words, each SPE can execute two (different) instructions simultaneously. Instead of using a hierarchy of caches, each SPE has a local memory for audio, video and image processing.

We have introduced three types of Multicore architectures: Multicore homogeneous processors, Multicore heterogeneous processors, and Multicore SoC. In this book, depending on the addressed topic, we use either a specific definition, or the generic term Multicore when concepts are applicable to all solutions. In fact, the following Section depicts a possible future scenario of convergence of Multicore architectures into a common paradigm.

1.1.1 Trends in Multicore Architectures

As discussed in the previous Section, Multicore solutions, such as computing platforms for PC and embedded systems, rapidly prevail. This Section briefly analyzes important technology, market and application issues that allow understanding current and future trends in the evolution of Multicore architectures. Execution parallelism has been the common denominator in the evolution of processor architectures, especially coping with reduced memory access latency. Homogeneous multicore is a good example. Adequate application parallelization and programming models are required to follow and exploit this trend. PC market devices demand application-agnostic architectures, while embedded market also becomes open, at least in terms of pre- and post-silicon configurability. On the other hand, high performance and general purpose processors tend to merge in energy efficient single die solutions. Multicore era is evolving towards a common architecture template based on heterogeneity and flexibility. Multicore communication between processor and off-chip memory has an access time of several hundreds of cycles. During this time the processor is stalling, degrading overall performance. For this reason, several research activities on innovative hierarchies have tried to cope with this problem. Larger, sophisticated caches exploit temporal and processor locality, and thus limit expensive memory access time (e.g. remote accesses are reduced by a factor of 10 with a hit ratio of 90%) at the expense of increased chip area; indeed, most of the silicon area in homogeneous Multicore processors is occupied by caches. Notice that temporal locality occurs since memory accesses tend to refer to recently referenced words, while processor locality occurs since memory accesses tend to refer the same processor.

Another approach to cope with memory access exploits application parallelism at different levels and granularities. Processor evolution has moved away from bit level parallelism which increases the word size of the instruction to ILP (Instruction Level Parallelism) which focuses on parallel execution of multiple instructions, leading to architectures based on pipelining, out-of-order and speculation techniques, superscalar execution, or VLIW (Very Long Instruction Word). In superscalar systems, a hardware dispatcher distributes multiple instructions to multiple hardware functions, while in VLIWs, hardware has been simplified and instruction parallelism has been explicitly decided at software level.

Another strategy that improves performance is simultaneous multithreading (called multiple contexts or thread-level parallelism). This technique, similar to multiprocessing, provides a way to overlap computation and communication through parallel execution of relatively independent sequences of instructions (called threads) occurring between context switches. While a simple superscalar issues multiple instructions from a single thread during every cycle, symmetric multithreading issues multiple instructions from multiple threads during every cycle. Thus, multithreading increases concurrency by keeping the processor busy executing tasks, while waiting for inevitable off-chip memory access delays. Alewife is an example of this architecture that has been designed in the late 1980s [71]. Medium-grain multithreading boosts processor throughput and execution unit efficiency, despite increasing memory latency, while fine-grain multithreading (or multiprocessing) exploits either embarrassingly parallel or data parallel applications, or coarse-grain ILP arising from independent problems. Usually a small number of contexts (e.g. two or three) may be sufficient for achieving most of the performance gain. Thus, updating applications from single to multiple threads can increase performance, provided that enough memory bandwidth is available to satisfy data access requirements.

A major drawback of multithreading is scalability and performance. Multithreading cannot benefit much from systems with a single cache, single integer arithmetic, and a single floating point unit. Multicore is a viable alternative to simultaneous multithreading, since scalability depends on the number of cores available in the die which can be easily increased. Moreover, multicore is the only way to scale performance at a reasonable power budget. However, since scalability depends on application parallelism, besides reducing the external memory gap and memory contention due to simultaneous access, maintaining the required concurrency for keeping a Multicore processor busy is also a critical issue. As a trivial example, it is possible to assume a Multicore processor with 8 cores, with the ability to execute 2 operations per cycle per core, i.e. up to 16 operations per cycle. With a cache miss every 8 accesses, there are (16/8) or 2 misses on average per cycle; assuming an average off-chip memory access time of 100 cycles, up to 200 independent tasks are necessary to fully utilize the Multicore processor.

Nowadays, homogeneous Multicore processor solutions in the PC market replicate usually large optimized cores in systems of two, four, eight cores. However, the research and industrial multicore community seems to trace a roadmap towards heterogeneous multicore computing. The historical Inmos Transputer and the recent Cell from IBM, Toshiba and Sony are examples of this trend of integrating general purpose processor architectures with specialized processor templates. Another example is the announced AMD Fusion product which combines both CPU and GPU. These are the first steps, but looking into the future, a huge number of simpler and smaller cores with a more specialized instruction set architecture (ISA) will be integrated on a single die. Since the number of cores in a Multicore processor is rapidly increasing, a Multicore processor requires more and more parallelism to be fully utilized [20].

Current desktop applications or Microsoft Windows operating system functions do not easily parallelize, but the large number of heterogeneous applications executing on a PC may be able to provide the right degree of concurrency. In the future, we anticipate a major interest in updating software from single to multiple threads for increasing performance.

Moreover, application parallelization and hardware support must progress together. In particular, available memory bandwidth can be a severe bottleneck in multicore systems, demanding substantial paradigm shift. Thus, for some application types, instead of processor-to-memory communication, direct processor-to-processor communication provides a viable solution towards supporting high concurrency; an example is streaming-based engines. In fact, efficient design of a specialized multiprocessing system requires close correspondence between the hardware model of a processor and the software model of a task. This means that inter-task communication model must be compatible with the communication model among processors.

Nowadays, in both PC market and embedded domain products, on-chip interconnect solutions are based on buses or crossbars, which are becoming rapidly inadequate to support core-to-core communications. Moreover, wire delays and power start to dominate compared to gate. Hence, design of efficient on-chip interconnect architectures is the next real challenge. On-chip interconnect (Network-on-Chip or NoC) based on short point-to-point communication removes or reduces these bottlenecks, making Multicore architecture more scalable.

Current Multicore SoC solutions already exploit parallelism, distributing the application workload among cores or hardware accelerators, with a significant impact on the way to write applications. This is possible, since applications are known at design time, hence allocation (mapping) is static which can be better optimized. In Multicore processors, this approach is no longer valid, since applications are not known at design time, thus architectures must be open. Nowadays, this solution starts to become an option also for the embedded market. Although Multicore SoC can keep on exploiting application knowledge at design time, open configurable architectures eventually become mandatory. Therefore, configurable hardware/software architectures must be able to target several product derivatives based on the same silicon (different standards in different market regions) and guarantee reuse of the same architectural platform and IPs across different product generations.

Configurability can be performed by Original Equipment Manufacturers (OEM) companies instantiating or removing software components, or even by end-users downloading, installing, or removing software applications. It can also be performed at design time by semiconductor companies, e.g. by instantiating or removing system hardware parts or specializing IPs, such as RISC or DSP cores. Moreover, at the heart of communication-centric Multicore architectures, a key component in which configurability will play an important role is the network-on-chip. Configurable NoC will allow semiconductor companies to compete on product differentiation, while using standard fixed processor cores, such as ARM, and MIPS. This implies a competitive advantage against competition. As a paradox, instead of selling products with IP brands (e.g. ARM), semiconductor companies could brand directly architectures, promoting differentiation. Customization of consumer products is an important innovation for future embedded architectures. It is the enabling factor for a platform to scale to future products, reducing R&D investments and supporting future market requirements in the entire convergence space of computing, communication and consumer electronic applications.

Sometimes, when the number of end-users in a particular market is quite large, a possibility to address the cost of building such platform is by commoditization. For example, Google phone attempts commoditization of the mobile communications market (measuring currently 2.7 billion customers) by building a mobile platform together with the associated ecosystem (similar to what happened to the PC market several years ago). Google has announced an open software stack (called Android) to build the phone, and an Open Handset Alliance to overcome existing barriers to innovative devices and services. Chip manufacturers participating in this Alliance will build platforms with open technology, e.g. open register sets. This will lead to new commoditized Multicore architectures with high degree of configurability in which the number of applications and type of the application is left to the end-user, alike the PC. If this eventually happens, Multicore SoC for mobile phone market will follow Multicore processor business models, in which software will have an important role in product success. For instance, this has already been the case with many music players, such as Apple iPod, Microsoft Zune, Creative Zen, and Sony Walkman.

Although future Multicore architectures must exploit application parallelism in order to utilize efficiently available computational capacity, there is currently no clear vision of the future Multicore programming model; a short description of existing approaches is provided in Section 1.2.4.

For example, domain- or application-specific data parallel programming models form a promising solution that increases the degree of parallelism. By focusing on common data movements, memory access follows well-defined, predictable patterns. These patterns can then be exploited by suitable high-level programming abstractions based on a set of efficient hardware communication primitives to improve efficiency. Many wireless, Digital Signal Processing (DSP), and especially multimedia workloads that dominate the personal computer market are based on large-scale, block-oriented data parallel computations on input streams of digital signals. These applications are similar to data parallel scientific applications with special requirements, such as real-time constraints, and limited lifetime of data streams. Streaming is a natural approach to parallelizing most media applications. Stream processing does not map well to traditional Multicore processor architecture with caches, since it requires a constant thrashing of cache while requesting new data elements. These algorithmic requirements can be met efficiently by specialized cores or hardware acceleration components.

Streaming architectures provide a semi-regular fabric based on simple programmable processing elements with a general-purpose instruction set and local instructions and data, interconnected via a relatively simple network of point-to-point channels. The streaming engine addresses data-flow dominated applications, as well as intensive semi-regular tasks with deterministic real-time behavior. For example, an image filtering pipeline performs image acquisition, raster scanning, color conversion, convolution filtering of different types, then color conversion again and picture formation tasks. The input to (or output from) the pipeline can come either from memory, or from peripherals.

Streaming systems must take advantage of dynamic reconfigurability and regularity in software development and integration through existing and new programming languages, library extensions, and tools. Parallel programming languages mainly target reducing the required programming effort to express a parallel algorithm, while it may also offer performance benefits by avoiding sequential code, e.g. in executing rendering loops in a video game. In addition to sequential language extensions, research on general-purpose streaming languages includes Stanford's general-purpose Brook stream language compiler [50] and StreamC language used in the Imagine stream processor [162], University of Waterloo's special-purpose Sh library [212], and MIT's Streamit language which is based on simple filter, split, pipeline and join semantic principles [338]. Commercial tool kits also exist, such as SPI's Stream Processor's RapiDev tool suite focused on data parallel operations [339].

Digital convergence of functionality and user experiences has several implications to processor and SoC design, since it leads to new architectures able to provide high quality content handling to the end-user. Product manufacturability faces technology-related scalability issues, especially for power and memory bandwidth, while from a business perspective, time-to-market reduces today's development cycle to follow the quick turnaround time of each product generation. To address these challenges, we envision innovative software and hardware architectures that would provide enhanced end-user experience by connecting together through an advanced on-chip network, general purpose cores and specialized, configurable cores with a small local memory tha exploits locality.

We anticipate that NoC will evolve towards a programmable component that we define as Interconnect Processing Unit (IPU). These architectures will form a heterogeneous combinations of cores, including the IPU component properly configured and programmed. Similar to CPU, GPU, and other specialized cores that are based on a customizable (at design time) ISA depending on the specific application requirements, the IPU relies on a customizable set of services (SSA) dpending on system requirements. The IPU will expose a simple abstraction layer and programming model to applications, possibly enabling runtime configuration. The Spidergon STNoC IPU technology described in Chapter 4 is a first attempt at designing an IPU for advanced multicore platforms. In order to be successful, parallel software should be ready on time, and as mentioned in several roundtable presentations, it is imperative that the software doubles the amount of parallelism it can address every two years (a sort of Moore law for software) [45]. If this happens, architectural convergence between PC and embedded consumer market is likely, with both Multicore SoC and Multicore processors based on open configurable heterogeneous platforms.

In the following Section we present commercial and research Multicore solutions that may be considered as basic building blocks for a future envisioned platform.

1.1.2   Examples of Multicore architectures

Today there is a proliferation of commercial solutions providing innovative Multicore architectures configured with tens or hundreds of simple processor, memory, control and co-processor cores.

Unlike typical homogeneous supercomputing systems which focus more on grand challenge problems, such multicore systems feature high latency, improved network bandwidth, and lightweight communication and synchronization routines. These systems are used in modest coarse- or fine-grain parallel processing applications, achieving an order of magnitude better performance scaling, and reduced complexity and power consumption over traditional optimized general-purpose uniprocessor solutions due to limited speedup from instruction-level parallelism. In fact, multicore architecture based on interacting numerically-intensive tasks is very efficient for DSP and multimedia problems.

In our vision, such solutions aim to replace high-end DSPs and dedicated IPs in data-intensive mobile media processing applications of current Multicore SoC. For example, in Figure 1.1, tailored programmable cores may be replaced by multicore architecture solutions. These advanced products will become an integral part of several devices, such as cell phones, video processors, gaming stations and PDAs, integrating different application services, e.g. a cell phone might support high quality TV, online gaming and speech recognition capability.

Due to lack of detailed architectural and implementation data and use of different technologies, the rest of the Section provides just a brief description of existing and proposed multicore systems designed by major commercial manufacturers and research groups.

The PicoArray PC202 [105] packs 248 application-specific long instruction word DSP components (196 standard, 50 for optimized memory access, and 2 optimized for control, including memory access and multiplication); notice that PC101 packs up to 322 processors, including 14 function accelerators. Each component provides on-chip SRAM. Components are interconnected together and with a general-purpose processor through a custom mesh-like topology implemented with bus switches based on time division multiplexing (TDM). The picoArray chip targets low power, high performance communication systems. including cellular telephony, wireless, as well as complex digital signal processing applications involving stream- and block-based functionality, such as digital signal filtering, multiple access, FEC, equalization, synchronization. PicoArray programmers write software using an ANSI C compiler or high-level assembler for achieving code compactness and efficiency through code replication. They must also specify application partitioning (static communication flows) onto multiple chip processors using structural VHDL. The complete tool chain includes a C compiler, an assembler, a VHDL parser, a cycle-accurate simulator, debuggers, design partitioning, placement and routing and verification tools.

The Ambric AM series [11] propose a massively parallel architecture composed from 96 up to 360 clusters of 32-bit RISC processors with local registers and SRAM memory interconnected with asynchronous channels. Ambric can design a chip with virtually any number of processors in this range, while each cluster can run at a variable clock speed, thus matching performance to the workload. The Am2045 is designed to replace high-end embedded processors, DSPs, and FPGAs in applications that require fast general-purpose integer data processing and digital signal processing, including digital video compression/decompression. Ambric introduces high-level software development tools based on the open-source Eclipse framework. These tools statically compile a strict subset of single-threaded object-oriented Java applications into a native machine language, and automatically map the compiled objects onto the massively parallel array. Parallelism is expressed not through Java's standard thread class, but instead multiple instantiations of the same class run simultaneously on the parallel-processor array. To bind these objects together into parallel communication structures, programmers use either a graphical interface with a structural view of the program, or a textual language called aStruct. For increased performance, Ambric programmers can also directly code machine language objects.

The Connex CA1024 [78] is a multicore proposed initially for cost-effective consumer HDTV, based on streaming, audio processing, video encoding, decoding and transcoding. The multicore uses an efficient data parallel

ConnexArray architecture configured as a 2-d ring of (at least) 1024 RISC processors with local memory. The multicore supports a simplified vector instruction set of 70 instructions, including 16-bit integer and Boolean operations; no multiply-accumulate or floating point instructions are provided. In addition, the Connex programming language extends C with vector data types for expressing parallel operations on digital video data streams, such as vector operations.

The Cisco CRS-1 [101] carrier routing system is offering the industry's first true OC-768 bandwidth links which deliver 40Gbps connections. By connecting together up to 188 32-bit Tensilica RISC processors in a blocking multistage network that processes, queues and forwards incoming data, this system can scale from 1.2Tb/s, to over 90Tbps, far surpassing the next fastest IP router on the market. CRS-1 include high-level system tools for application programming, e.g. XML-based, possibly interactive interfaces for visual database management, routing policy selection, offline and online rollback-enabled configuration, proactive system monitoring, and role-based management.

3Dlabs [356] develops a fully programmable highly parallel, floating point array processor with 24 32-bit processing elements, dual ARM processor cores with multilevel caches, and tightly integrated peripheral support. This DMS-02 media processor is capable of delivering at optimum mips/mW rates a rich mix of media-intensive tasks, such as 2-d and 3-d graphics, accelerated streaming, e.g. video, audio, and imaging, and 16-bit floating point capabilities.

The IntellaSys SEAforth-24A is an extremely versatile user-defined embedded parallel array consisting of 24 mesh-connected 18-bit wide processors with local RAM and ROM and a powerful set of I/O functions [164]. Processors communicate data, status signals, and even code blocks. The array targets high performance and low power consumer applications, especially audio and video processing. The system supports VentureForth, a relatively complex, but compact RISC-based vector processing language with 30 powerful stack-, memory-, branch-, logic and arithmetic-oriented instructions. IntellaSys also supports Forthlets code that can be moved from core to core for specialized processing.

The Tilera multiprocessor [357] consists of 64 identical processor cores (called tiles) interconnected with Tilera's iMesh on-chip network. Each tile is a complete full-featured processor, including integrated L1 and L2 cache, while a non-blocking switch that connects the tile to the mesh. Each tile can independently run a full operating system, or multiple tiles can run the SMP Linux multiprocessing operating system, thus supporting an immense body of common open source tools. The integrated Eclipse-based Tilera multicore development environment features an ANSI- C compiler, a full-system simulation model, flexible command-line interfaces, and innovative graphics tools that help programmers code, debug, and profile large-scale multicore applications. Moreover, the parallel library iLib supports efficient inter-core communication, including socket-like streaming channels, lightweight message passing and shared memory.

Boston Circuits provides a Multicore processor solution with 8 to 16 cores for complex user applications and media processing targeting mainly electronic consumer devices, such as IPTV media centers, network display, and printers [48]. The proposed architecture implements a scalable, high-speed on-chip network that connects together 32-bit RISC cores and system peripherals, such as PCI and USB. This solution integrates a smart memory controller that handles common operations, such as memory copies. In addition, an on-chip "time machine" simplifies application development by handling scheduling, resource allocation, communication, and synchronization.

The XMOS multiprocessor is being developed by David May, the main architect of the INMOS transputer and the author of the Occam parallel programming language based on send/receive primitives in the 1990s; Occam is essentially derived from Hoare's CSP parallel programming model which is used extensively for automated reasoning of parallel programs [38]. XMOS consists of software configurable families of 2-d processor arrays based on the event-driven, multithreaded 32-bit XCore processor. It is not clear yet whether XMOS will be general-purpose, or address a specific application domain or narrow set of applications, such as communication base stations and switching fabrics. Nevertheless, it appears that XMOS will improve programming complexity, real-time response, power consumption and silicon cost over FPGA solutions. In particular, XMOS will use an embedded software approach with conventional, concurrency-lacking C/C++ programming with multithreading provided from the newly proposed XC language.

In 2006, the **Intel Teraflops Research Chip** prototype has been built within its visionary Tera-scale Computing Research Program which investigates the hardware and software technology necessary for building processors with tens or even hundreds of cores in the next 5-10 years. More specifically, Teraflop chip involves several innovations, such as advanced circuit design techniques, rapid tile-based design, 2-d message passing-based network-on-chip, power-efficient architecture based on mesochronous clocking, sleep transistors, and clock gating, and advanced fine-grain power management based on the required application performance. The Teraflop prototype chip connects 80 simple cores with 20MB of SRAM operating at 3.1 GHz, each containing two programmable floating point engines in an 8×10 2-d mesh configuration. This chip is the world's first programmable multicore processor to deliver 1.0Tflops/sec performance and 1.6Tbytes/sec aggregate core-to-memory communication bandwidth, while dissipating only 62W. Its power consumption is comparable to desktop processors and exhibits a scale of four reduction over the ASCI Red, the world's first general-purpose computer to benchmark at a teraflop in 1996.

The **University of California at Irvine asynchronous array** of simple processors [393] supports multiprocessing for DSP applications. The array is composed of simple processors with local memory and FIFO-based communication configured as a 2-d mesh.

The Stanford Smart Memories project [323] (SMM) focuses on hardware primitives and reconfigurable mechanisms required to implement a universal computing- and/or memory-based multicore environment for probabilistic reasoning algorithms used in application domains, such as data mining, image analysis, robotics, and genetics, as well as multimedia and DSP applications, such as ray tracing with global illumination in real-time, and speech or voice recognition. Thus, SMM integrates within a multicore a coarse-grain reconfigurable memory system (with programmable wires and logic) which supports diverse programming models, such as cache coherent shared memory (Single Program Multiple Data, SPMD) with speculative multithreading, streaming for data parallel applications, and transaction coherence and consistency [348] (TCC): the goal is to allow application-specific architectures to be programmed in the model that provides the best performance or programming ease. In particular, SMM supports simple cache coherent shared memory implemented using a conventional cache coherency protocol (MESI), and streaming based on optimizing automated static mapping computation kernels to processor cores. Based on the stream program and a machine model, Reservoir Labs' R-Stream [292] high-level cross-compiler generates Stream Virtual Machine code [196] (SVM), i.e. target-specific ANSI C language code augmented with high-level mapping instructions. This code is subsequently compiled using the low-level Tensilica XCC compiler. Moreover, SMM supports also parallel TCC code executed as simple transactions, including communication and coherence. This model automatically handles synchronization, without programmer intervention. At transaction completion, shared memory is updated atomically, and processors that have speculatively read from modified data are restarted.

1.2   Complexity in Embedded Software

A typical Multicore SoC design is becoming increasingly software-intensive due to multiplatform design, real-time performance, robustness, and critical safety constraints. It is typically estimated that 40% of the design time-to-market is spent on application code, 40% on hardware, and an additional 20% on making software work on top of hardware.

Figure 1.3 shows the software stack of a complex SoC, indicating the tremendous effort required in software development of Multicore SoC, which is unable to keep up with exponentially increasing system-level complexity and growing number of client services. For instance, in software terms a set-top box (or digital audio processing system) software stack requires more than 1 million lines of code.

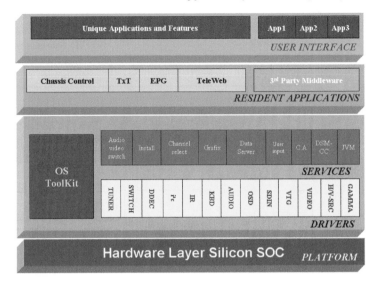

FIGURE 1.3: A complex software stack

For instance, in hardware terms designing a recordable DVD (or hard disk drive) requires more than 100 engineer years, while in software terms a set-top box (or digital audio processing system) software stack requires more than 1 million lines of code.

Multicore SoC design becomes software-intensive with improved performance, greater functionality and safety, relying on multiprocessing, multithreading, prefetching, cache coherence, and consistency aspects. Moreover, embedded software is expected to run on diverse hardware platforms and different operating systems. Thus, in order to reduce time between consecutive products versions, significant architectural changes, constant support calls to the original software team and duplication of testing efforts should be avoided. For this reason, an increasing number of companies target multiplatform software development in a reuse-oriented way through abstraction. As shown in Figure 1.4, a 4-layered efficient Multicore SoC design approach tries to decouple applications from low-level kernel space, including real-time operating system, system libraries, device drivers, and hardware platform.

The bottom layer consists of hardware and firmware components (processor and peripherals) providing functionalities, such as instruction sets, timers, memory and peripheral access to the software layers. It is often useful to abstract physical interfaces into logical ones. For instance, a PCI bus can be exposed to upper layers through generic point-to-point or multipoint bus APIs [337].

The next layer consists of the real-time operating system (RTOS), the system libraries and the driver debug port (JTAG). Notice that implementing

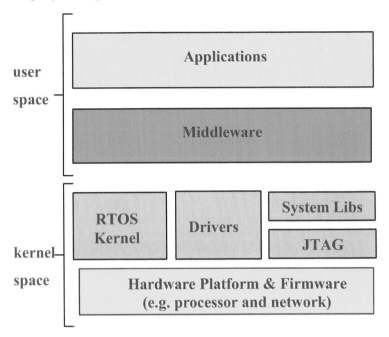

FIGURE 1.4:   SoC architectural levels

wrapper functions with a standard abstract functionality irrespective of the underlying layer helps in migrating and scaling applications and test benches to different platforms supporting a common RTOS model or abstract communication interface, e.g. the MPI message passing library.

Middleware is a software layer that sits on top (or instead) of the operating system, allowing developers to provide distributed services without considering low-level driver or operating system issues. Middleware simplifies complex application coding, providing portability across hardware and operating systems.

Finally, the top layer consists of application software that is independent of the underlying hardware platform. Furthermore, specialized libraries of reusable software IP components implement the necessary functions for developing systems in specific application domains.

Software components based on hardware-independent parts can be easily reused through intermediate standardization. Thus, industry standard APIs (high-level libraries) exist between applications, middleware and RTOS. Since reusable hardware-dependent software is currently impossible, VSIA has created a working group to propose a new industry standard for a hardware-dependent layer that will isolate hardware from software, simplify access to hardware components, offer scalability and enable reuse for distributed systems [310].

The following sections provide an overview of Multicore software layers with a top-down approach.

1.2.1   Application Layer

The precise application types that homogeneous and heterogeneous Multicore processors will execute are not known at design time and their complexity increases year after year. The end-user who buys Multicore processors loads and configures essentially all system and application programs necessary for a specific use case. For this reason, Multicore processors are generally open systems.

The situation is completely different for devices based on Multicore SoC which target tight market requirements. Typical examples are the set-top-box and HDTV systems which are sold directly to the public with all relevant application software. Similar scenarios exist in mobile phones. However, for certain consumer devices the situation is changing. For example, in smart phone and phone game applications, although a major part of application software may have already been uploaded, end-users may provide additional features.

During recent years and probably even more in future evolutions of consumer applications, multimedia content has driven an unprecedented revolution in the entertainment business, where MySpace, Youtube, Google are well-known examples.

Nowadays, we can distinguish among professional, personal and communication contents. Professional contents are delivered using set-top box via pay TV, Internet portals, mobile services, Internet search engines, such as Google, Yahoo, and illegal sharing, such as eMule and torrents. Personal contents refer to different ways to create and share information, such as MySpace and Youtube. Communication content refers to common end-user services, such as voice email, SMS, MMS, and blog.

All these contents are associated to complex, heterogeneous application software executed on different devices, such as PC, video games, mobile phones, and portable music and video players or in client server way. For example, typical applications in a mobile terminal are word processing, e-mail, web-browsers, powerpoint, java, digital media viewers, photo viewers, and Digital Rights Management. With software design complexity increasing with time, application designers must utilize advanced middleware or RTOS layer services.

1.2.2   Middleware Layer

Applications running different types of software, or operating on different computer platforms often need help communicating with each other just as people who speak different languages often require a translator to communicate.

Middleware refers to communication software that connects together distributed embedded application programs or software components coded in different programming languages and operating on different heterogeneous multicore platforms, so that they can exchange data. Hence, middleware hides the hardware and OS from the application layer, abstracts complex hardware and software functions and provides a natural environment for application development which promotes reusable components. Distributed middleware computation includes telephony, application servers, knowledge management systems, multimedia processing, web services, mobile environments, and Internet.

Middleware is distinguished into general-purpose and application-specific systems that support proprietary, open standards or combinations.

General-purpose middleware typically integrates an application manager, whether a virtual machine (e.g. Java Virtual Machine), a component object model (e.g. COM or DCOM) or a container architecture (e.g. CORBA), along with the interactive engine, libraries and databases.

Application-specific middleware is often provided by the device manufacturer. For example, ST Nomadik multimedia framework provides a programming model and an associated environment for simplified development of mobile phone applications. In addition, application-specific middleware based on open standards implemented by many industry players (e.g. DAVIC, TV-Anytime, MHEG for multimedia presentation, and MHP for interactive TV) is licensed to set-top-box manufacturers for analog, digital, cable, or satellite pay TV (e.g. MediaHighway by French Canal+, Microsoft TV, NDS Core, OpenCable in US Cable market, and OpenTV Core).

1.2.3   RTOS – Drivers - System Programs

An RTOS is a special-purpose operating system with the necessary features to support embedded (firm or hard) real-time applications whose correctness depends not only on the correctness of the logical result of the computation, but also on its delivery time. Explicit and implicit timing constraints are derived in the requirements phase by examining the physical environment. In extreme real-time systems, an RTOS may also have reliability, robustness, availability and safety constraints. Notice that reliability is the rate of system failures, while safety is the degree in which system functions do not lead to accidents causing loss. Reliability does not imply safety and vice versa.

An RTOS consists of several modules, such as system tables, scheduler, communication, synchronization, and interrupt-service routines. System tables provide:

- a task descriptor for running each task,

- a device descriptor for using each I/O device, and

- service descriptors specifying parameters for RTOS requests.

The RTOS scheduler (dispatcher) determines which process will run next after a time-out, or when a blocked process triggers a context switch to a new task, after saving and restoring all necessary info. Processes alternate in a ready, running or suspended state. For each of these states there is usually a queue containing the corresponding process descriptor. Since processes are broken into threads, scheduler operations include thread create, run, suspend and exit operations, as well as synchronization primitives, e.g. mutex lock and unlock. RTOS scheduling algorithms must allow each task to execute on time, if this is possible. Several scheduling algorithms have been proposed, such as

- polled-loop, i.e. round-robin rotating between all tasks,

- phase- or state-driven based on discrete finite state machine states,

- interrupt-driven based on priorities,

- multitasking or parallel environments exploiting program locality,

- preemptive schemes that context switch upon high priority requests, and

- hybrid schemes based on a combination of these algorithms.

The RTOS includes various communication and synchronization primitives. Locks provide for mutual exclusion, while semaphores describe both mutual exclusion and scheduling constraints. The semaphores update atomically a counter indicating the number of system resources, thus providing a simple synchronization and mutual exclusion mechanism for several problems, such as producer/consumer synchronization with bounded buffers.

Other synchronization primitives include conditional variables which offer mutual exclusion based on predicate calculus. Event flags allow a task to wait for a single condition or a combination of conditions represented as bits in a vector sequence. Therefore, a consumer may wait for a producer to signal a necessary condition prior to consuming data. Event flags may have broadcast semantics, i.e. when a task sets some flag bits, all tasks whose requirements are now satisfied are woken up one by one in priority. Finally, signals provide asynchronous event processing and exception handling. Communication primitives are based on queues, i.e. message queues and pipes for multiple messages, or mailboxes for single messages.

An Interrupt Service Routine (ISR) runs on the processor for each I/O device and may trigger a system interrupt. The ISR controls the I/O device, providing data transfer between the device and the RTOS. High-priority ISRs are executed quickly.

RTOS choice usually depends on the selected processor core and application domain, e.g. Symbian OS for portable wireless devices, OSEK for automotive systems and PalmOS for PDAs. Modern RTOS provides fast, predictable behavior with respect to timing (by providing formulas for context switch, reschedule, page faults, pipelined execution, synchronization, and interrupt handling), dynamic process priorities (established during runtime instead of compile or link time), preemptive scheduling, device driver support, graceful error handling and recovery, and tools for analyzing real-time scheduling using Rate Monotonic Analysis (RMA) techniques.

Commercial RTOS systems have become widely popular, with the market growing at some 35 percent each year.

RT Mach supports predictable and reliable firm real-time UNIX processes and real-time synchronization [359]. RT Mach supports static process priorities with earliest deadline first scheduling, rate monotonic scheduling and priority inheritance protocols.

CMU's Chimera has a priority based, multitasking microkernel with real-time scheduling based on fixed or dynamic priorities, with rate monotonic scheduling (RMS) and earliest deadline, or maximum-urgency-first algorithms [336]; RMS is based on task rates, e.g. the task with the highest rate has the highest priority, while RMA analyzes timing constraints using external tools offline. Memory has no inter-task or intra-task protection, no paging and no virtual memory. Devices have Unix-like abstractions. Chimera is highly predictable, with fixed context switch time, dynamic reschedule and semaphore non-blocking calls. It is robust to processor exceptions and memory corruption, and provides error signaling.

Mentor Graphics Corp. is making VRTX, the real-time executive OS SoC version, available through a community-source model [285].

For systems with multiple CPUs, parallel operating systems are based on simple modifications of the Unix OS, e.g. CMU's Mach OS. Symmetric microkernels execute on each processor, providing graceful degradation for faults. Each microkernel includes libraries for inter-process or external communications, I/O, interrupt handling, process and memory management. In addition, it provides support for parallel languages, interactive/batch execution, security, and accounting. The latest parallel systems provide, along with computing nodes, redundant (fault tolerance-related) computing nodes, as well as dedicated OS and I/O nodes.

Device drivers provide a well-defined API between hardware and software. They control setup and operation of an IP block, e.g. initialization, runtime configuration, testing, enable/disable function, and I/O processing. Driver operation is based on events that capture interrupts sent from IP blocks. Device drivers are called frequently with their response time being critical to overall Multicore SoC performance. They also increase IP reuse and enable multiplatform software development at reduced cost and risk.

A common approach to designing device drivers for multicore SoC is to partition driver functionality into different levels.

Low-level drivers provide basic functionality and are generally not re-entrant. They exploit the available communication and synchronization protocols that on-chip interconnects will offer for exchanging information between application and peripherals. For example, in the latest products based on ST Micro-electronics' STBus interconnect designers can decide to transfer information in different modes, using a simple read or write, a set of consecutive reads or writes, or a pipelined transfer.

High-level device drivers are abstract and have complex functionality. They use low-level drivers for accessing the hardware. These drivers are re-entrant, since they interact with the application. They must be able to handle peripheral activity, while the application is running.

Today's multiprocessor applications have diverse computation, communication and I/O requirements leading to a wide range of sought-after QoS criteria, e.g. minimum network bandwidth for streaming applications, and maximum latency, jitter, or packet loss. While our demand and dependence on software grows, our ability to produce it in a timely, cost efficient way continues to decline.

The rest of the Section discusses software design problems in Multicore SoC, focusing on parallel programming paradigms, concurrency, and consistency.

1.2.4   Multicore Programming Paradigms

While Multicore SoC is attractive from power and performance perspectives, it introduces the need for simple and efficient parallel programming. Alike the shift to assembly language and object oriented programming in the 1950s and 1990s, multicores require a parallel operating system along with a user-friendly parallel programming model [169]. These programming models are high-level abstractions of the Multicore architecture used by programmers to code applications. These paradigms define coordination of the Multicore architecture, specifying the manner in which different software components collaborate. Moreover, programming models specify communication and synchronization operations.

Multicore SoC or general heterogeneous multicore programming is more difficult than parallel programming, since it is domain or application-specific, and thus heterogeneous.

The heterogeneous multicore programming model must support a highly parallel instruction set with advanced features, like data parallel processing, streaming of large and complex data structures, real-time discrete-event processing, and application mapping that enable fast development and increase portability. However, automated parallelization is an exacting task, since parallel programming cannot be abstracted or layered, while code and data must be structured in non-intuitive communication patterns to take advantage of special hardware features, such as remote transactions, or DMA. Moreover, parallel programs for multicores are hard to debug, due to hazards, such as combinatorial explosion of execution interleaving, deadlock and data race conditions occurring upon simultaneous data access from multiple cores or threads.

1.2.4.1   Flynn's Taxonomy

In 1966, Flynn provided an interesting four-way classification of parallel architectures into SISD, SIMD, MISD and MIMD, based on a division between program control and data [85] [193] [103]. It is interesting and challenging at the same time that all four programming models may simultaneously occur in modern day multicore SoC and heterogeneous multicore programming.

Single-instruction single-data stream (SISD) corresponds to the traditional Von Neumann "stored program" sequential computer.

Single-instruction multiple-data (SIMD) enables a centralized data parallel approach, where all participating processing units execute the same instruction on multiple sets of data and synchronize before executing the next instruction. In addition, most SIMD (hybrid) architectures offer fast scalar instructions through deeply pipelined units. SIMD is frequently called single-program multiple-data (SPMD). In SPMD, some processors may be selected to abstain from a particular type of operation. SIMD architectures include versions of modern general-purpose processors, e.g. Intel's Pentium and Motorola's PowerPC, traditional parallel systems, such as the TMC CM-2, and DSP architectures.

SIMD processing is applied to vector processors and SIMD arrays. Vector processor exploit fine-grain data parallelism by performing many operations with a single instruction and are appropriate for multimedia and scientific applications, e.g. image processing or linear algebraic calculations for weather forecasting. In contrast, SIMD arrays integrate seamlessly computationally intense data parallelism with scalar processing for common applications, such as real-time DSP digital signal filtering or decompression, where small amounts of scalar code cannot be decomposed into vector operations and may lead to dramatically lower speedups.

Multiple-instruction single-data (MISD) exploits temporal ILP, by setting pipeline stages and executing several independent instructions simultaneously, e.g. vector pipelining in Cray-1. Pipelining has inherent limitations caused by hazards in dispatch logic and unavailability of ILP for creating a continuous stream of similar operations. Superscalar design with multiple (beyond two) instruction issue units have rapidly declining speedups and waste transistors, if limited ILP exists in application programs. Very Long-Instruction Word (VLIW) systems were devised to reduce the complexity of issuing multiple instructions simultaneously. However, VLIWs are often more complex than superscalar systems of similar performance, suffering since compilers cannot efficiently schedule resources due to insufficient runtime information, e.g. branch predictors in non-continuous data-streaming applications. VLIW (called also EPIC, for explicitly parallel instruction computing) was implemented in high-performance processors, such as Texas Instruments' C62x DSPs and Intel's Itanium and Merced.

Unlike pipelining, superscalar and VLIW techniques that exploit ILP using complex temporal or spatial means, coarse grain process- and thread-level multithreading based on the natural independence of processes or threads, and fine grain data parallel vector processing are transistor intensive techniques, relying on array data path elements rather than complex control structures. Thus, multiple-instruction multiple-data (MIMD) exploits spatial instruction-level parallelism (ILP, or control parallelism), where processors may execute different instructions. Typical MIMD systems are the BBN Butterfly, Intel Paragon XP/S, Cray T3E, and the current top performance supercomputer, NEC Earth Simulator which achieves 35 Tflops/sec. MIMD systems can also support SIMD capabilities. Two types of multithreading can be identified: continuous and simultaneous.

Continuous multithreading exploits medium-grain thread-level parallelism within a single instruction stream to keep the execution pipeline busy by switching to a new thread when a thread stalls, e.g. during remote memory access. This new thread is an instruction stream that is data, memory, and control independent of other streams. Multithreading languages include: C-like Cilk supporting process spawn, abort and blocking or nonblocking synchronization [74], Split-C providing efficient access to a global address space [329], and Guava (Java extension) that avoids hazards [23].

Fine-grain simultaneous multithreading (or hyper threading) interleaves instructions from several threads onto the machine's execution units for near real-time response [362]. Thread-level parallelism avoids latencies due to pipeline hazards. Moreover, while thread programming simply spawns multiple independent processes that manage atomic transactions efficiently for embarrassingly parallel applications, such as application and web servers, in SoC applications it is often vague compared to other approaches, such as data parallel or message passing-based programming. Thus, in general, for simplicity and predictability reasons, threads are not considered appropriate for SoC programming.

1.2.4.2   Existing Multicore Programming Models

Recently, there has been increased interest in applying traditional parallel programming on complex multicore architectures. In addition to well-defined parallel programming models, experts commonly distinguish between implicit and explicit programming.

Implicit parallel programming uses sequential algorithms and a "smart" compiler for parallelizing an application. Implementations using implicit parallelism are usually simple but not as efficient as explicit ones, since parallelism is often not too easy to expose. Since it is often extremely difficult for a multicore, parallel system compiler to efficiently parallelize a sequential algorithm, it is usually better to manually redesign a data parallel algorithm. Unfortunately, few Multicore SoC programmers are trained enough to design such cost-effective solutions.

In explicit parallel programming the programmer must specify exactly how sequential processes are created and how they cooperate. Exposing application parallelism for multiprocessors is not always easy. Several programming paradigms are used, such as general-purpose message passing, data parallel and (virtual) shared memory models, or domain- and application-specific programming models supported through high-level languages and analysis or transformation tools.

Explicit parallel programming paradigms differ in fundamental aspects, such as the time programmers invest in writing parallel programs, the way code distribution and data allocation are specified, the architecture for which the paradigm fits best, and the maximum amount of application concurrency that can be exploited.

In message passing, processes update local variables and exchange information using explicit send/receive operations. Hardware platforms naturally suited for this model are multicomputers with point-to-point connections, multicores and distributed clusters. Message passing is easy to emulate on other platforms. The message passing programming paradigm is based on sequential languages augmented with standard message passing libraries, such as PVM and MPI. Notice that MPI versions exist for real-time (MPI-RT) or fault-tolerant computing (MPI-FT, or MPICH-V).

Data parallel programming supports SIMD or SPMD control flow abstraction. Data parallelism is expressed by assigning data to virtual processors, which perform identical computation on its data. Data parallelism may involve not only arrays of numbers, e.g. for dense linear algebra computations, but also application-specific parallel symbolic computation, searching, or sorting operations on sets, multi-sets, maps, and sequences, as specified in STAPL [333], SETL [303], or NESL [42]. Several existing sequential programming language compilers, e.g. Fortran90, High Performance Fortran (HPF) or C* [209] support general-purpose data parallel execution by inserting complex parallel operations that reduce programming complexity, improve performance, and establish correctness by avoiding deadlock or data race hazards.

Currently, the commercial RapidMind development and runtime platform enables parallelizes single threaded applications fully across multicores, managing also its execution [283]. In addition to general-purpose extensions, modern platforms provide SIMD instruction set extensions, such as vector load, add or store, such as 128-bit wide AltiVec instructions for PowerPC [94] [95]. Open source data parallel extensions are also available [213] [318].

In shared memory programming, processes read and write directly from/to a shared address space. The programmer has to express code distribution only, since most data are stored in global memory. Since data can be accessed concurrently by different processes, mutual exclusion operations are vital. Shared memory languages consist of sequential languages, together with special libraries providing memory access and synchronization subroutines. For example parts of MPI-2, Open-MP, and ShMem are shared memory programming utilities.

In Virtual Shared Memory (VSM) a global virtual address space is shared among loosely coupled processing nodes. This helps programmers and hardware designers in development of data and computation intensive application. Virtual shared memory may be implemented through parallel communication libraries in higher layers of the network protocol stack. Although distributed memory is viewed as a shared address space, the VSM programmer has to bear in mind that remote memory access may significantly increase memory latency, and thus he must try to exploit spatial, temporal and processor locality. Virtual shared memory languages are similar to shared memory languages. Nowadays most commercial supercomputers are NUMA VSM systems.

An alternative to classical lock-based concurrent programming is nonblocking synchronization, which provides mutual exclusion without using locks, thus improving performance at the cost of increased programming effort [225] [201] [142] [145]. A common solution encapsulates nonblocking protocols in library primitives, but this limits generality with which nonblocking programming can be employed by ordinary programmers.

Efficient development of reliable, portable, high-performance parallel software using message-passing and especially concurrent shared memory or nonblocking synchronization is difficult even for expert programmers; for example, locks are either used too often, harming performance, or too little, harming correctness by causing hazards, such as deadlocks or data races. Hence, new high productivity multicore languages must reduce the gap between parallel and mainstream application development programming languages, raise the level of abstraction, and provide support for efficient prototyping and production quality code through additional tools, such as debuggers and performance profilers.

A promising approach stemming from theoretical research on programming models almost two decades ago is the transactional memory. This approach focuses on coordinating parallelism and providing mutual synchronization in Multicore, multithreaded or highly parallel shared memory systems through a general and flexible way that allows programs to read and modify memory locations atomically as a single operation, similar to a database transaction [143] [134] [144] [146]. If no dependencies are detected between transactions at runtime, then parallel transactions can execute their own load and store operations atomically. If dependencies exist, then certain transactions restart, and corresponding load and store operations are serialized. Since transactions are of arbitrary size and duration, the programmer specifies only where a transaction begins and where it ends, making parallel programming accessible to the average programmer. Software transactional memory supports atomicity by extending programming languages, compilers, and libraries, while hardware transactional memory supports atomicity through architectural concepts, e.g. content-addressable memory.

1.2.5 Concurrency

Concurrency can be classified in two principal categories. Pseudo concurrency is equivalent to multithreading on a single processor, while actual concurrency involves many processors and/or processes with instructions executing simultaneously. While sequential algorithms impose a total order on the execution, parallel algorithms only specify a partial order. Consequently, a parallel algorithm is equivalent to two or more algorithms each specifying part of the final result. Concurrency can dramatically increase real-time performance by increasing the degree of simultaneity. However, parallel programs are complex to design, test, integrate and maintain, since major issues concerning resource management are introduced.

To ensure correct application functionality the system must be free of deadlock. A deadlock may be defined as a cyclic dependency of not granted resource requests for buffer or channel resources. Deadlocks may occur when processes mutually exclude each other from obtaining resources from a shared pool of resources.

Starvation refers to tasks repeatedly being denied resources. Starvation usually reflects problems of fairness in static allocation or dynamic scheduling policies, either in the processor or system interconnect.

Data races arise when threads or parallel processes access shared variables and at least one access is a write operation, while there is no synchronization as to the order of accesses to that variable.

Priority inversion occurs when a lower priority task holds a resource required by a higher priority task that waits on the same resource. If the system enters fail-safe mode, priority inversion may lead to timeouts causing continuous system resets.

Deadlock, starvation, data races and priority inversion pose functional risks to parallel programming in real-time systems, including data consistency and system failure, since debugging can be very frustrating and costly. Thus, new tools that identify the sources and evaluate the effect of priority inversion are very important in scheduling analysis. Existing program analysis tools, such as "Eraser" that detects deadlock and data races, are major steps in eliminating these problems [300]. In respect to deadlock, two techniques review this problem from different angles: detection and recovery, or prevention.

Deadlock detection and recovery can be based on resource dependency graphs. This method is simple and effective only if the state of the resources and tasks is available.

Runtime deadlock prevention based on spin locks is the most effective strategy, but in its simplest form it offers no guarantee of fairness and has an increased overhead due to loss of flexibility, resulting from a static or adaptive prevention strategy. A spin lock ensures that only one processor may modify a critical section in a shared data structure at any given time. Spin locks execute enormous number of times in concurrent data structures, e.g. in RTOS priority queues, or fault tolerance recovery techniques. These costs are alleviated in modern processors and RTOS using distributed synchronization mechanisms, such as Ticket lock and MCS lock that prioritize requests, remove central access mechanisms and allow for efficient resource management [230].

In terms of performance, hierarchical locking refers to the ability to lock small parts of concurrent data structures, or even program code, e.g. critical sections, for exclusive access, thus increasing the level of concurrency. Lock granularity is an important tradeoff between increased concurrency and design complexity, e.g. in database access one may hierarchically lock the complete database, particular tables, or data elements [130].

1.2.6   Consistency

A memory consistency model must be specified for every level at which an interface is defined between the programmer and the system, eliminating the gap between expected program and actual system behavior. There are techniques that prevent, or detect and recover from this possible incoherence.

For a single von Neumann processor, program order implies that for any memory location, a read routine to that location will return the value most recently written to it. However, for a multiprocessor, we must specify the order in which memory operations may be observed by other processors. Thus, either implicit (e.g. in an SIMD system) or explicit synchronization (e.g. via semaphores or barriers) may be necessary when a new value is computed and other processes access this value.

The use of atomic hardware primitives, such as test&set, fetch&increment, fetch&add, fetch&store, compare&swap and load-linked/store-conditional with the additional guarantee that all pending memory operations are completed before performing an atomic operation, provide several potential advantages. In fact, concurrent protocols are easier to code, code is easier to understand and debug, and an efficient implementation is possible.

The simplest consistency model for shared memory programming is sequential consistency (SC). SC was defined as follows. A multiprocessor is defined as sequentially consistent if the result of any execution is the same as if the operation of all the processors were executed in some sequential order, and the operations of each individual processor appear in the order specified by its program [202].

Although SC provides a simple and intuitive model, it disallows some hardware and compiler optimizations possible in uniprocessor architectures, e.g. prefetching or nonblocking write, thus enforcing a strict order among shared memory operations [148]. In practice, we encounter many relaxed memory models, as in IBM 370, Total Store Ordering (TSO), Partial Store Ordering (PSO), Processor Consistency (PC), Weak Ordering (WO), Release Consistency (RC), PowerPC, Relaxed Memory Order (RMO) and Alpha consistency [3]. With relaxed memory models, the programmer or the compiler must provide synchronization operations enforcing a consistent memory view, e.g. by placing a memory fence (barrier) after every memory operation, or more intelligently by looking for potential access cycles.

However, use of aggressively relaxed consistency models is arguable, since with the advent of speculative execution, these models do not give a sufficient performance boost to justify exposing their complexity to low-level software authors [148]. Even without instruction reorders, at least three compiler-based optimization methods exist: prefetching, multithreading, and caching. Data prefetching based on look-ahead or long cache lines exploit spatial locality, i.e. memory accesses that refer to nearby memory words, thus reducing application latency. Unlike multithreading, prefetching relies on the compiler or application user to insert explicit prefetch instructions for data cache lines that would otherwise cause a miss.

1.3   Technological Issues in Multicore Architectures

Technological issues challenge traditional SoC performance and reliability. Thus, Multicore SoC design methodology must adapt to the deep submicron effects, such as growing noise sensitivity caused by supply voltage reduction, small voltage swings, higher manufacturing tolerances and increased importance of capacitive and inductive crosstalk. Considering submicron effects and complexity issues, synchronization and power consumption demands for energy efficient design techniques at all different levels of product development are real challenging. Silicon complexity increases, and for instance, at 45nm a design can count 1 million gates per mm^2. The high costs associated with generating a mask in modern fabrication facilities, and the decreasing time-to-market window for leading edge products, place extreme pressure on design teams to produce high yield and first-time correct silicon. Multicore solutions have to be manufacturability-aware, to receive all benefits from the integration capability of nano dimensions, which evolve according to Moore's law.

This Section makes a quick overview of main challenges for multicore architecture manufacturability from the perspective of advanced CMOS technologies.

1.3.1   Deep Submicron Effects

As explained in the International Technology Roadmap on Semiconductors [167] and the European EDA Roadmap [107], several new challenges emerge in silicon integration which relates to manufacturing technology.

Non-uniformity and atomic-scale design effects in next generation CMOS processes, cause signal fluctuations in gate tunneling current, threshold voltage and gate-length critical dimensions. Thus, for example, a metal wire not only connects functional elements, but also depending on its material and geometry introduces resistance, fringing, or parallel plate capacitance and parasitic inductance, i.e. electromigration effects.

The parasitic effects introduce random design defects that are catastrophic, since they change implementation characteristics and electrical behavior of devices and wires; indeed, they reduce operating margins (supply voltage, temperature, and signal-to-noise ratio), clock frequency, and ultimately design performance, increasing delay and energy consumption. In addition, the introduction of noise due to electromigration and electrostatic discharge gives rise to signal integrity and system reliability issues.

As microelectronics technology continues to reduce the minimum feature size, the gap between the masks in sub-wavelength dimensions and what is really manufactured on silicon is widening significantly, due to new sources of variation. Consequently, predicted model performance at design level may significantly differ from the results obtained after silicon manufacturing. Since variability reduction through process control requires innovative process control and tuning techniques during fabrication and expensive manufacturing equipment, the impact of parameter variations should be compensated with advanced design solutions and tools. For example, a major benefit comes from taking the right decisions early in the device process design flow (TCAD), perhaps even at system level, thus drastically reducing the number of iterations before tape-out. Thus, coping with variations early in the design has potentially significant advantages in terms of shrinking time-to-market and reducing costs for process control.

Following technology scaling, although process variations steadily shrink in absolute terms, they are actually growing as a percentage of the increasingly smaller geometries. Moreover, variability sources grow in number as the design process becomes more complex, and correlations between different sources of variation and quality metrics of the design process are becoming more difficult to predict. Process parameter variations usually impact the design. Variability can often cause catastrophic yield loss, i.e. manufactured chips do not function correctly. In other cases variation introduces parametric yield loss, i.e. manufactured chips do not perform according to specification (chips may be slower or consume more power than expected during the design). In sorted designs, such as microprocessors, parametric degradation means that fewer chips end up in the high-performance, high-profit bin. In other design styles, such as ASICs, circuits below a performance threshold must be thrown away.

Obviously, the first kind of yield loss has traditionally received more attention. Typical chip functional failure is caused by the deposition of excess metal linking wires that were not supposed to be connected (bridging faults), or by the extra etching of metal, leading to opens. Several techniques address catastrophic yield loss, including critical area minimization, redundant via-insertion, and wire bending/spacing. Although for a long time parametric yield loss has been an overlooked problem, nowadays, it is becoming more and more important, since design performance can be dramatically affected by process variations of submicron technologies. For designs exclusively based on optimization of process parameters, the analysis may be inaccurate and synthesis may lead to wrong decisions, when these parameters deviate significantly from their nominal value. Despite recent research efforts, only few methods can address parametric yield loss, e.g. design centering and design for manufacturing (DFM).

The increased electrical resistance reduces voltage (IR drop) and causes extra delay in interconnect wires. The increased capacitance increases wire delay and the amount of operating power that the integrated circuit requires. Capacitance also causes crosstalk; this effect is due to the coupling capacitance between interconnections, a signal switching on a net list (aggressor) may affect the voltage waveform on a neighboring net list (victim), resulting in signal errors. The increased inductance is a high-frequency phenomenon that affects circuit operation for example causing inductive voltage drop (Ldi/dt noise) on the power distribution network and inductive coupling crosstalk.

Electromagnetic interference emissions (EMI) can severely and randomly affect analog and digital circuit functionality, causing adverse effects in signal integrity and supply coupling. For example, high frequency currents resulting from digital switching activity in the power supply network of an IC may leak out to the package or PCB causing EMI at the harmonics of the clock. This is particularly true, when the wavelength determined by the clock frequency is approximately the size of wire length or circuit dimension.

1.3.2   Power Consumption in CMOS Devices

Low power design techniques have been employed for more than 25 years, particularly in watch circuits and calculators. Nowadays, energy efficiency is the main design constraint in communication and control systems in everyday life or portable battery-powered consumer electronic appliances; new power-efficiency Mips (or Tflops) per mW metrics start to appear as main quality metric for multimedia terminal, laptop, digital cellular telephony and emerging applications, such as ambient intelligence and sensor networks. The immense interest in today's energy-efficient microelectronics systems stems from environmental laws, ethical reasons, energy delivery cost for stationary systems, extended battery life for portable systems with data intensive applications, and difficult thermal heat dissipation management [76] [138]. Power consumption does not scale with deep submicron technology (below 65nm), due to increase of clock frequency, high design complexity (in number of transistors) and increase of leakage power. Therefore, power profiling is needed to identify opportunities for optimizing power-critical parts for different applications, e.g. through built-in device-specific power management based on different operating voltages that may be offered by the technology.

System power is computed by summing the power consumed by all resources, interactions, and the environment. Consumed power can be classified into four components: dynamic, short-circuit, leakage and static.

Dynamic power can be partitioned into internal consumption by the cell and energy for driving the load, including wiring and fan-out capacitances.

Dynamic power consumption of CMOS circuits can be described as $P = 0.5\alpha f C V^2$, where f is the clock frequency, α is the switching activity that refers to bit transitions per clock cycle, i.e. charging and discharging of output capacitance, C is the load capacitance which is roughly proportional to the chip area, and V is the supply voltage. The total capacitance is the sum of the input capacitance of fan-out gates, wiring and parasitic capacitance. While f and V are directly defined by the designer, C is determined by the system architecture, and depends on application, mapping and architecture. Nowadays, ~80-90% of all power dissipated in a circuit is due to switching activity and 90% of this is due to dynamic power. Dynamic power is under control, by reducing f, α, C or V at transistor-, gate-, register transfer- or system-level. For example, at vendor process technology-level, load capacitance can be similarly reduced by using at the physical layer silicon on insulators junctions and low-k dielectrics constants. Similarly, at design level, it is possible to influence dynamic power dissipation by using new low power versions of CMOS technology, routing high-frequency signals on the least capacitive upper layers, increasing data parallelism and scaling clock frequency, and especially the supply voltage level [55]. However, since a reduced frequency extends program execution time and energy consumption is a product of execution latency and power consumption, scaling down frequency saves dynamic power but is ineffective in providing energy savings [64].

Short circuit power dissipation is caused by a short-circuit current that flows from power supply directly to ground during the nonzero interval between fall and rise time in CMOS circuits when both the p- and n-device of a gate are conducting, i.e. during switching activity [368]. This power is wasted and never collected by output capacitances. Dynamic and short-circuit power depend on signal switching activity, while static does not. Static power dissipation depends on current flow from power to ground during idle time. Static power is the product of the power supply voltage and a static current composed of leakage and sub-threshold components. Leakage currents, common in all bulk MOS devices, are due to parasitic diodes formed in the transistors. They are usually critical in battery-powered applications with long standby or sleep time. Leakage currents flow through the reverse biased diodes in gates between the diffusion regions and the substrate or through transistors that are non-conducting. All leakage power parameters are determined by technology, device dimensions and junction temperature. Sub-threshold currents depend on critical sizes, such as oxide thickness. These are orders of magnitude larger than the ones that occur in circuits designed with analog techniques or those that use resistive pull-up devices. Static power will soon be as high as dynamic power, since natural leakage of transistor is multiplied by 10 every 2 years. This is already true at high temperatures.

A number of effective techniques are available to the designer for minimizing dynamic power consumption, including technological, architectural, and algorithmic software methods.

Multiple design methodologies are available for reducing dynamic power by decreasing switching activity, e.g. by invoking smart data representation, data and resource allocation, programming styles and scheduling algorithms that minimize the number of basic data flow operations or increase correlation between successive input patterns of functional blocks and reduce overall signal switching activity. Most behavioral synthesis approaches are based on iterative improvement heuristics [63] [194] [297] [281]. Latency hiding techniques, e.g. using cache to explore data locality, multithreading or prefetching are also helpful in reducing power by minimizing global communication over long wires with high capacitance load [227].

Static power can also be controlled using a number of complex techniques. Multi-threshold CMOS design (called power gating) scales down the supply and scales up threshold voltage to meet power vs. performance tradeoffs. Since scaling down threshold voltage, exponentially increases sub-threshold leakage, so-called sleep transistors with high threshold voltage are inserted to function units or gates. Sleep transistors are turned off during the sleep mode, which can significantly reduce the leakage. Moreover, multiple supply and threshold voltages takes advantage of the fact that independently optimizing different parts of the design is likely to result in different values for the optimum supply and threshold voltages for each macro-level block or standard cell (called voltage islands).

1.3.3   Clock Synchronization

Traditional design techniques are based on the assumption that all circuits are synchronous. Ideally, transitions are controlled by a clock signal that arrives simultaneously at all gates in all SoC computing resources; thus, signals reach their destination within a predetermined number of clock cycles. But clock is a major source of SoC noise, power dissipation and electromagnetic emission, since it synchronizes during every clock cycle millions of flip flops. As a result, although it is hard to accurately estimate clock power at system level due to possibly unavailable architectural and technological design parameters, clock tree power consumption can reach 40% of the total dynamic power in processor cores for current technology [309].

Technological trends, including low-k dielectrics, increase clock power dissipation. A single distributed clock leads to a synchronous digital circuit, but today this is hard or impossible to implement. Asynchronous design techniques, such as clock gating, frequency and voltage scaling, prevent switching in unused areas of the chip, resulting in significant clock power savings. For example, Globally Asynchronous Locally Synchronous (GALS) protocols in Multicore SoC architectures use several synchronous clock domains communicating over asynchronous channels to reduce total power.

Since standard design flows or tools are not available for asynchronous design, new design techniques and architectures resistant to variability become important.

System correctness and robustness are based on the clock period, since signals must obey certain inequalities (margins) in respect to timing. The clock cannot be distributed at the same instant to all on-chip components, because of the clock skew, defined as the maximum difference in clock arrival times (positive or negative phase shift) in any two periods, i.e. the clock signal in a synchronous circuit arrives at different components at different times. Clock skew is caused by unavoidable variations in buffer load, interconnect wire lengths, manufacturing process spread across die changing resistance, inductance, capacitance values, or temperature gradients. While performance of digital circuits is generally more immune to mismatch than analog circuits, synchronous design is sensitive to skew variations in the clock distribution network. Clock skew contribution to clock cycle time can be reduced by minimizing total wire length using an H-tree (or other possibly hybrid structures) with non-uniformly placed intermediate dummy buffers. Jitter slightly changes the clock period due to variations in Phased Lock Loop (PLL) oscillation frequency and noise sources. Jitter can be reduced by minimizing power supply noise (IR and Ldi/dt). Moreover, jitter due to PLL noise also improves with CMOS scaling.

1.3.4   Supply Power

Power/ground integrity becomes a serious challenge, since supply voltage fluctuations may translate to timing inaccuracy or even result in signal integrity problems that may cause the circuit to fail. Technology scaling imposes requirements on the reduction of power supply voltage. Digital circuits induce load charging and short circuit currents into the power supply network at the clock frequency and each harmonic. IR drop causes wavelength-related resonance, supply voltage and consequently delay variations within digital blocks that may produce severe signal integrity problems. Substrate potentials especially affect analog components in mixed-signal systems, since supply noise can easily couple to the analog part that shares the same supply and a common substrate. The most obvious solution is to reduce the maximum distance between supply pins and circuit supply connections, i.e. using a finger-shaped power distribution network. Design and layout of the power distribution network must be performed early in the design process and later gradually refined.

1.3.5   Fundamental Concepts for Efficient SoC Design

For increased productivity, profitability and market share, important challenges in silicon and system-level design complexity must be addressed. Although research is at an early stage, it is speculated that new Multicore SoC platforms in emerging advanced nanometer technologies (below 45nm) will be based on the integration of ideas spanning across several fields, including new interconnect or device materials, improved manufacturing processes based on statistical modeling and variability analysis, advanced computer architecture, operating system, parallel and distributed computing, and innovative EDA methodology and tools. This Section outlines fundamental concepts for reliable, high-performance silicon design: Design For Manufacturing (DFM) which improves yield and reduces design uncertainty, exploitation of IP reuse based on synthesized IP assemblies, use of multi-level regularity paradigms, and hierarchical heterogeneous SoC design to simplify validation, verification and optimization efforts. Notice that the last three concepts apply to any level of abstraction.

1.3.5.1   Design for Manufacturing

Technological advances in deep submicron lead to an exponentially rising transistor count, but increase silicon complexity and introduce parasitic deep submicron effects, as explained before. In addition, system-level design complexity also increases, e.g. in platform-based SoC hardware/software co-design. The widening productivity gap between increased silicon/system design complexity and EDA language and tool capacity (RTL or system-level) implies that beyond technological solutions alone, innovative design approaches are necessary to bridge the gap with design complexity and maintain an increase in functionality with reduced cost and time-to-market constraints.

Operating variations caused by noise, power supply, temperature and manufacturing process variability due to parasitic effects in lot-to-lot (within a fabrication lab), wafer-to-wafer (within a lot), global inter-die (within a wafer) and local intra-die (device-to-device) significantly impact design reliability and performance specifications. Canonical circuit examples assess the impact of new technology generations on process variability, including dopant levels, scattering or diffusion processes, and layout distribution characteristics of defects.

Existing low-level, SPICE-based EDA tools and technology models based on rule checking can model parametric yield loss and defects by introducing a statistical design of intra-die variability based on (pessimistic) worst or extreme case behavior, i.e. the same sensitivity to variations is assumed for all circuits. However, the implementation with design or model uncertainties often leads to circuit over-design and inaccurate performance estimation. To account for DSM variability, a new phase (called parasitic effect extraction) has been introduced in the design flow after post-layout synthesis. This phase involves extensive simulation with computationally-intense probability density functions describing variations of all technology parameters. In addition, since limited information is exchanged between the actual design and manufacturing processes causing a logical- and physical-level dichotomy, higher-level design tools could allow automated model analysis and optimization that efficiently achieves time closure of the design.

Traditional EDA design flow methodologies and system architectures focus on abstraction levels from functional to transaction level modeling (untimed, with timing or cycle-accurate) to RTL behavioral and structural simulation and finally gate and transistor level simulation (after placement and route). All levels are neither redundant nor resilient to defects, while increased and disorderly system complexity makes things even worse.

1.3.5.2 IP Reuse

IP design reuse is the ultimate means to bridge the technology gap. It is a hot topic in application-specific SoC development, including digital, analog, mixed-signal IP design and verification [21].

Efficient application of reuse methods of hardware and software IP blocks early in the design process allows SoC designers to develop application-specific virtual system prototypes using pre-built silicon functions with desirable, consistent and efficient properties, thus drastically decreasing time-to-market, risk and costs. Complex SoC design is usually based on integrating primitive design elements through high-level models of hardware IP, e.g. USB, processor, bus and memory controller, and software blocks, e.g. application software, device drivers, communication stack, and RTOS. System-level reuse ignores implementation details, simplifying specification, plug and play integration at different abstraction levels, hardware/software co-design, validation and verification, while enabling efficient design space exploration.

The introduction of reuse establishes an evolution in design flow methodology that requires new techniques and best practices that are briefly discussed next.

Bottom-up component-based design supports IP reuse and modularity with multiple interoperable abstraction levels, protocol refinement (separation of behavior and interface), compatibility testing and automated verification.

Top-down platform-based design (PBD) is an integration, verification and interoperability-oriented approach emphasizing systematic IP reuse for developing complex products based on kernel components, such as processor cores, RTOS, interconnects, memory and standard interfaces, as well as qualified embedded software. PBD aims to reduce development risks, costs and time-to-market. Along with fast and easy integration of IPs, PBD provides a continuous and complete design flow from hardware/software partitioning to RTL generation.

High-level synthesis (HLS) refers to automated hardware/RTOS synthesis derived from high-level IP modeling. HLS can be based on hierarchical modeling and automated generation of interfaces satisfying by construction system properties.

While IP reuse enhances design quality by selecting pre-validated components, model-driven design (MDD) focuses on exploring design alternatives at a conceptual level. MDD promotes creation of a series of models from initial problem specification to final implementation and integration, defining relationships that define which model and what abstraction level must be used at a given time. Consequently, MDD allows architects, designers and developers to build larger, more accurate, and maintainable software-intensive systems, enhancing interoperability and verification. This design methodology separates business and application models from underlying platform technology. For example, executable Unified Model Language (UML) is expected to raise the abstraction level, e.g. in component or platform independent models. Notice that platform-specific models provide interfaces, services or facilities that can be generated automatically from the base model.

More idyllic IP reuse developments are expected by converging work on rapid prototyping, model-driven design, language compilers, automatic code generation, graph rewriting, verification (code coverage, constraint satisfaction and model checking), meta-modeling and ontology engineering. Besides technical issues in IP reuse, business models and legal positions are also vital.

1.3.5.3   Hierarchical System Design

Hierarchy is a fundamental system- and/or block-level property that enables efficient design of complex systems of organized complexity based on two different, usually disguised design methods: a top-down analytical approach, i.e. the divide and conquer paradigm, or a bottom-up synthesis approach, i.e. the design-reuse paradigm. Behavioral hierarchy enables top-down design, i.e. modeling of system behavior using sequential (hierarchical finite state machine-, procedure- or recursion-oriented) and concurrent (parallel or pipelined) decomposition. Structural hierarchy enables bottom-up design, i.e. the realization of new components based on existing ones; thus, a system is represented as a set of interconnected components at different abstraction levels. Hierarchical SoC design focuses on the decomposition of conceptual and physical processes found in complex (control, memory, network, and test bench) modules into simpler, subordinate modules controlled by efficient high-level system models, with clean, modular interfaces.

1.3.5.4   Design Regularity

To limit the exponential complexity of global design methods, there is an increasing interest in subdividing global problems into locally decoupled problems, and then assembling a large number of similar, local solutions through scalable and compositional transformations. Multi-level symmetry (or design regularity) is a concept applied to physical, logical and architectural levels of a complex SoC design. This concept is based on a limited number of simple, modular patterns that extremely simplify fundamental design, similar to "keep it simple" design in early RISC processor design. Regular cell-based design must be exploited at different levels of abstractions to improve flexibility and design productivity, or increase yield by reducing design uncertainty from silicon. Common methods that help increase regularity involve network-on-chip design focus on static and dynamic mapping, new semiconductor devices, such as processor-in-memory, and high-level synthesis tools

For example, regularity of the network-on-chip infrastructure and processors simplifies the process of partitioning and mapping common applications with (semi-) regular communication or memory access pattern patterns, such as scientific or multimedia applications involving intensive communications, such as broadcast, all-to-all communication onto the architecture. A similar way to increase regularity in network-on-chip design is through topology reconfiguration based on dynamic embedding, and predictable characterization of electrical and physical properties, e.g. by injecting randomness to communication design, or by considering structured communication channels, hierarchical clock methodologies, or more uniform power distribution.

Regular fabrics explore synchronous or asynchronous multicore distributed memory architectures and global hierarchical on-chip communication. New high-level regular semiconductor devices will allow faster system-level sign-off and direct path to physical implementation. For example, processor-in-memory (PIM) is a high-level architecture that seamlessly logic merges CMOS logic with DRAM memory blocks on a single die for greatly improved regularity, and consequently higher manufacturing defect tolerance, and fast and direct memory access for next generation operating frequencies [166] [272]. In this effort, logic patterns, implementation platforms, communication algorithms and computation paradigms must become even more regular. Thus, in PIM, processors are simple execution units, e.g. Arithmetic Logic Units (ALUs) or DSPs which provide a dynamically controlled granularity that delivers requested performance and reconfigurability. PIM can implement different interconnect topologies, enabling high reliability through active fault tolerance and multithreading schemes.

For several process generations, classical CMOS scaling enabled rapid shrinking of feature size by aggressively scaling the wavelength of light. This technique has improved ASIC performance, while increasing versatility and decreasing cost at the same time. The immense complexity of nanometer technologies introduces new challenges for the synthesis of complex SoC designs that affect performance, power, functionality, yield and manufacturability.

Multicore circuits are often characterized by highly regular data path-intensive structures, such as bit-slices corresponding to arithmetic or multiplexing operations on wide buses, and interconnect building blocks among replicated bit-slices. This is especially true for multimedia and DSP applications, such as image processing, signal processing and graphics. Although algorithms and methods for extracting and utilizing design regularity in data path compilers have not been extensively studied yet, it is obvious that regularity must be preserved for resolution enhancement techniques (RET)-compliant lithography-friendly designs in advanced process technologies.

Therefore, in order to make the physical layout functional, manufacturable, and compact, conventional high-level logic synthesis methodology must be extended to seamlessly detect and extract regular logic patterns at the functional, structural, or topological layout level during top-down or bottom-up design flow, and integrate them among layout optimization objectives [252] [289] [379]. This is crucial for decreasing wire length, improving timing, and reducing wiring congestion. It is also important for increasing productivity by achieving timing closure, facilitating fabrication and decreasing manufacturing cost by reducing the raw number and complexity of design rules in RET [182] [172], and increasing system-level layout predictability and yield by controlling better deep sub-micron process variations.

More specifically, current regularity-aware design methodology focuses on implementing regularity at the physical level through restrictive design rules [214], layouts on grid [376], or by using a limited set of lithography-friendly patterns and circuit topologies [173]. In the latter case, design flow begins with high-level RTL description and a set of design constraints. A minimum set of required logic components necessary to meet design specifications is created by the "brick discovery" step, before converting RTL into a regular brick-level netlist. Finally, during placement and routing, the brick-level netlist is used to create the final physical implementation, e.g. as a 2-d arrangement of bricks and achieve high levels of manufacturability and yield.

Chapter 2

On-Chip Bus vs. Network-on-Chip

2.1 Transition from On-Chip Bus to Network-on-Chip

SoC communication is currently realized using traditional on-chip buses, such as ARM's AMBA AHB or AXI [17], ST Microelectronics' STBus [301], IBM's Core-connect [160], and the Sonics Backplane [327]. On-chip buses are not fundamentally different than computer buses, except that they are designed and optimized to operate entirely within a single chip, thus wider buses are possible and there is no constraint to the number of wires. These buses typically include address, control and data wires, bi-directional signaling for management, and complex arbitration.

Buses are well understood and have been successfully implemented in many complex SoCs focusing on both application-based and processor-specific features, thus providing compatibility with most available IP and processor cores, and SoC communication requirements.

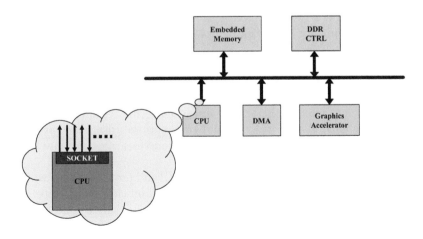

FIGURE 2.1: Example of an on-chip bus

On-chip buses provide a set of input and output (I/O) wires (called socket interface) and define a common inter-core communication language known as bus protocol. Sockets handle bus protocol and traffic requirements. Since in on-chip buses socket interface is included in the core, there is no distinction between socket protocol and bus protocol. Having a common standardized bus or socket protocol allows easy plug and play core integration without the need of a redesigned socket. An example of an on-chip bus is shown in Figure 2.1.

We generally identify two types of sockets: master and slave. While master is associated to an active component which initiates the communication protocol, slave corresponds to a passive component which essentially responds to master requests. The shared bus is usually based on a centralized address decoder that drives a single multiplexer which selects the master, but sends signals to all connected (slave) devices using high fan-out logic, propagating address and write (or read) data signals. Then, the selected slave responds to the master's request by setting read data signals (or acknowledgments) that eventually reach the master IP. Timing for all on-chip bus signals is provided through a single clock shared by all master and slave cores attached to the on-chip bus. Besides channels and buffer memory, a key element in the definition of a bus is arbitration. This process responds to master requests by granting access to the bus and driving address and read/write data signals to a slave after decoding the destination of the transaction. If the slave is unable to respond immediately, then the transfer is either delayed, or the transaction is retried at a later time.

Buses reduce cost by sharing wires and implementing a well-defined communication protocol that allows simple addition or exchange of new IPs or processor cores among systems that use the same bus standard.

However, buses require complex arbitration, especially when supporting devices with widely varying latencies or data transfer rates. To improve performance with slow cores, advanced split operation mode enables a slave to delay the operation for later completion, thus releasing the bus for other transactions.

An increase in the number of connected IP and processor cores causes a bus communication bottleneck due to sharing of aggregate bandwidth by all attached units, and worsens arbiter delay, thus introducing variation, limiting scalability and making the arbiter instance-specific. Additional concerns regarding reachable clock frequency and time closure arise, since every bus transaction involves all connected cores and unnecessarily consumes significant power.

SoC design trends towards a large number of IP and processor cores with different types of application traffic drive evolution of on-chip bus architectures. In the typical system scenario of a single slave device, where concurrency is not required, the main issue is the reachable operating frequency for a given number of IP cores. In a multi slave scenario, the aggregate bandwidth is limited, since concurrency among transfers is architecturally impossible.

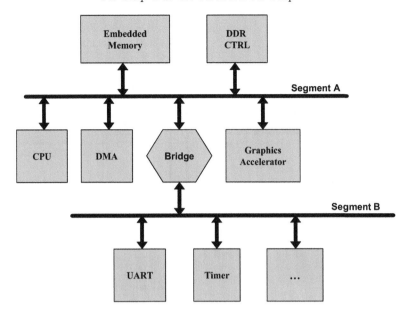

FIGURE 2.2:   Example of segmented bus

New multiple hierarchical (called also segmented) on-chip bus structures with several bus segments have been introduced to improve concurrency and reduce contention and power consumption. An example is shown in Figure 2.2. To avoid signal degradation, data are pipelined over long, segmented wires, while possibly expensive bridges are usually needed for transferring information among different bus segments. In typical applications, bridges use two ports to connect segments of different data word size, while they sometimes perform protocol conversion. When all buses use the same protocol, bridges can be implemented as simple fifos. In this case, each bus segment can be finely tuned to achieve latency and frequency requirements.

Segmented buses reduce capacitance and power, since they limit the number of direct connections to the bus, i.e the number of connected cores. However, distributed bus arbitration corresponds to the aggregate action of multiple arbiters, thus it is often very complex and time consuming to compute optimal overall settings. Moreover, segmentation increases latency and requires a lot of upfront partitioning. When transfers are larger than the bus width, they take several clock cycles, thus the bus remains unavailable and this situation creates more latency.

In order to increase network bandwidth and reduce latency, multiple parallel buses have also been introduced. These buses allow simultaneous point-to-point master-slave connections, thus acting as a crossbar. Crossbars interconnect master to slave cores in any possible combination. As a consequence, network bandwidth increases at the expense of extra complexity, decoding and arbitration time.

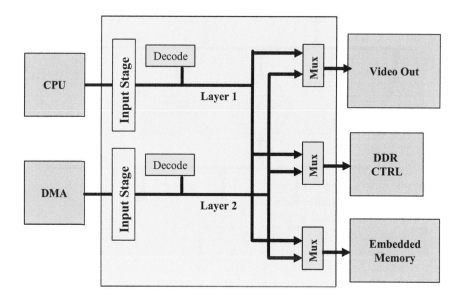

FIGURE 2.3:   ARM's multilayer AHB

Figure 2.3 illustrates a representative example for parallel buses based on ARM's AMBA AHB multilayer architecture. This can be derived from Figure 2.1, when the CPU and DMA master cores are connected to three slave cores: the IP core, the DDR memory controller, and the embedded memory. First, the input stage stores data until they are transferred. The decode stage determines which slave is chosen for the transfer. A multiplexer routes the transfer between the correct master/slave combination. If CPU and DMA want to access external memory (or other slaves) simultaneously, then arbitration is performed at the slave side.

Although advanced features offer pipelining or retry techniques, traditional bus protocols are still based on unified transactions, i.e. the request and response phases of the data transfer are coupled to each other. Requirements for high performance led to the introduction of multiple outstanding, exclusive, out of order and above all split transactions, such as in OCP2.0 and AMBA AXI standards. As described in Chapter 4, with split transactions, the request and response phases are completely decoupled and normally implemented on top of two independent paths: the request and the response path. This separation has been introduced to avoid deadlock hazards that the communication protocol may introduce.

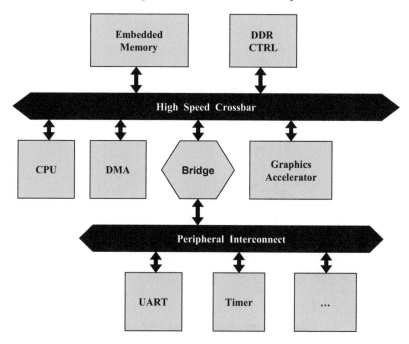

FIGURE 2.4: High-level abstraction of a multistage crossbar interconnect

The evolution from segmented buses to crossbar-based interconnects is illustrated in Figure 2.4. Large single stage crossbars are complex and expensive to implement, since the number of wires (or cross points) grows exponentially with the number of master and slave components. Since cost-performance metrics are important, in essence, a shared bus has limited scalability in terms of available bandwidth, while a crossbar does not scale well in regard to hardware complexity.

This is the main reason why multistage interconnects built by interconnecting multiple smaller crossbar switches in cascade through hardware fifos (called bridge in Figure 2.4) have been typical commercial solutions for most SoC architectures, especially for the high speed crossbar shown in Figure 2.5. The multistage crossbar represents a first attempt to offer the right degree of scalability and customizability required by the SoC. STBus developed by ST Microelectronics, IBM CoreConnect, and AMBA AXI interconnect solutions are some well-known examples.

Multistage interconnects based on small crossbar switches increase average delay with respect to a single crossbar, and in general, they are not appropriate for non-uniform traffic, since they are unable to exploit spatial locality of references. In addition, all control and data signals of an IP socket are routed in parallel through the multistage network from master to slave, resulting in limited wiring efficiency and possible backend routing congestion. Above all,

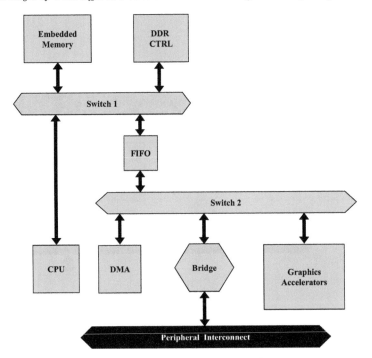

FIGURE 2.5:   Architecture of a multistage crossbar interconnect

bandwidth requirements for complex Multicore systems are pushing towards higher operating frequency that are not attainable by Multicore solutions. As technology advances, implementation of multistage crossbars does not scale, since deep submicron effects increase the resistance-capacitance delay of a typical wire. For this reason, pipelined, packetized multistage crossbars are being proposed for complex SoCs. Such solutions introduce several pipeline stages in the crossbar and multiplex data and control over a reduced set of wires to achieve higher operating frequency.

As shown in Figure 2.6, pipelined, packetized multistage crossbar solutions introduce a layered communication approach, similar to ISO/OSI stack. In contrast to the bus view, on-chip interconnects are not anymore seen as a passive component for plugging onto it several cores, but rather an active component that provides connectivity and services to cores. Thus, socket interface is not anymore included in cores, but it is implemented in new components, called network interface units (NIs). This provides a clear de-coupling between the external core protocol, e.g. an AMBA AXI, VSIA, or OCP and internal interconnect communication (among switches), also strengthening modularity, composition and multi-protocol interoperability of Multicore. Common crossbars lack additional services that pipelined, packetized multistage crossbars offer, such as security control, QoS, and power management. Arteris is

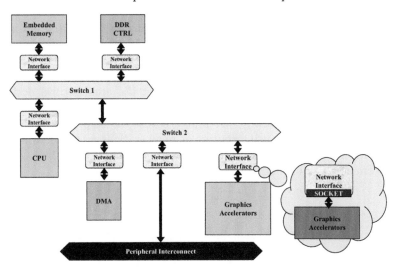

FIGURE 2.6: Pipelined packetized multistage crossbar

a relatively new company specializing in this type of technology, while the Sonics MX solution, although not packetized, is another very representative example.

Fundamental characteristics of new interconnects described above, such as packetization, enhanced services, and orthogonalization between socket protocol and transfer communication are key NoC design concepts. Nevertheless, in our vision, Multicore architectures will eventually require more disruptive solutions. Adoption of networking fundamentals within the on-chip domain must be pursued with an even more aggressive attitude, to design communication-centric platforms able to accommodate increasing application complexity, while maintaining efficiency in design productivity.

For complex Multicore products implemented using emerging advanced process geometries, especially 45nm, 32nm and 22 nm, electrical noise and crosstalk effects are becoming worse. In this scenario, on-chip delay on interconnection wires is a major limiting factor for performance, while managing silicon faults is another fundamental challenge for SoC design. As a historical consequence, distributed communication based on hierarchical multi-hop structures is no more an architectural option, but the only viable way to handle implementation limitations, while increasing performance and scalability.

Current interconnect technologies support cascades of multiple stages with a certain topological freedom, or switch pipelines, but these measures are not enough. Multicore demands a drastic shift towards distributed on-chip network infrastructures, built using a set of basic routers properly arranged in space, but perceived, designed and controlled as a structured element from the system perspective.

Although basic routers can be optimized, relying on the limited fan in/out (called indegree/outdegree or -arity) and predictable routing, to keep up with Multicore design complexity and demand for sophisticated services and enhanced performance, the interconnect cannot be anymore the architectural design of the router building block, but rather the architectural design of a simple, cost-efficient network based on a cascade of these switches. This network needs to be physically distributed across the chip, highly pipelined at the architectural level (using routers or relay stations rather than backend buffers), inherently redundant (with multi-path flexibility), and above all, able to perform as a flexible and controllable structured element.

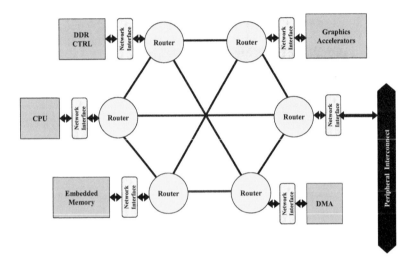

FIGURE 2.7:   SoC equipped with a Network-on-Chip

As shown in Figure 2.7, on-chip interconnects based on point-to-point network topologies can provide scalability and power efficiency. Topology is not only the way routers are connected, but it is a fundamental infrastructure for almost all computation, communication, and synchronization properties and architectural features. Thus, topology is a crucial ingredient behind the concept of structured element as described above.

In the research community, many proposed on-chip networks are essentially parallel computing topologies which suit a planar SoC implementation. For example, the recently proposed Intel Teraflops is essentially a mesh connecting up to 80 cores, while the Tilera processor is also based on a mesh-like topology. Several other examples are provided in Section 2.2.

In the embedded consumer market, Multicore heterogeneity demands for a NoC topology which is extremely flexible in terms of customization to application-specific requirements. Notice however that architectural exploration for evaluating alternative interconnect topologies can be time consuming, reducing design productivity. In fact, convergence to an optimum application-specific topology is difficult, even if a considerable investment in advanced design methodology and tools has been performed.

A more feasible strategy for the future of communication-centric Multicore architectures is rather than searching for an optimal topology to provide a configurable robust infrastructure able to meet application- or domain-specific performance requirements in a scalable way. Notice that scalability comes either within a product design cycle when facing changes in system requirements, such as incorporating new IPs or processor cores, handling unpredictable software impacts, or among different product generations in order to build a rich product portfolio on top of SoC platforms.

NoC allows connecting through a standardized network interface in a plug-and-play manner configurable and parametrized IP and processor cores, or even entire subsystems from internal company libraries or external vendors, thus meeting time-to-market pressures. As described in Chapter 1, due to product convergence within the electronic consumer market and increase in CMOS integration capability, Multicore architectures try to assemble complex clusters around a host controller processor, implementing complete application systems, such as graphics, imaging, video, and modem. Each such system with the complexity of previous SoCs will be integrated on top of an efficient communication platform and an appropriate memory hierarchy. NoC can also be the infrastructure at subsystem level, especially with the emergence of parallel multicore architectures that can implement multimedia or graphic applications and achieve the required energy and performance tradeoffs.

Apart from distribution, with the previously mentioned topological implications, NoC evolution with respect to current pipelined, packetized multistage crossbar interconnects must also be pursued into other directions.

At first, available interconnect solutions are not really based on the network concept of routing, but rather on address decoding. At each stage of the interconnect, a packet has to be routed to the proper output port of the switch through tables which associate a switch output port to an address range, i.e. the memory map address of the IP socket interface. The address corresponding to the memory location of the read/write operation is used to simultaneously decode the final destination and also the routing path through the network. On the other hand, routing and especially source-based routing is more power efficient than using complex and slow routing tables, and performance is better due to much simpler switch implementation that avoids complex and slow routing tables. With these simple schemes, the path is deterministically established at the injection point based on the packet destination. Moreover, routers have to check the routing information in the packet header and implement trivial logic to decode the output port at each stage. Thus, unlike address decoding, where each stage requires an explicit association between address and direction, network communication becomes an end-to-end property, i.e. translation between the memory address of the operation, and the network destination and also the selected route to this target, is performed once and for all.

Managing the increased network complexity with a large number of connected agents requires simplified routing path control that enables fault tolerant design and provides enhanced performance. Routing is one aspect of the layering approach that needs to be applied at each network level in a pragmatic, but at the same time radical way. The router is required to be agnostic of the end-to-end protocol, i.e. the socket IP information cannot be used inside the network, but only at the network entrance (ingress) and exit (egress) point. Thus, routers are generic enough to route packets, independently of their semantics. In addition, the approach of re-instantiating the same router for all network sizes increases modularity and implies substantial cost savings in the design and validation of complex NoCs, since portability and IP reuse across multiple designs is crucial for all semiconductor vendors.

Although most current interconnect technologies have been designed to deal with request and response types of packets, e.g. load/store, Multicore evolution will lead to coexistence of different paradigms. Thus, the NoC interconnect must be able to deal simultaneously with different packet semantics, such as streaming, enhanced end-to-end flow control or distributed QoS policies.

Another fundamental network aspect not integrated in available pipelined, packetized solutions is related to virtualization, often the most efficient (and convenient) way to guarantee non-interference in communications. In general, Multicore needs to accommodate several traffic classes, several protocol message types, and several services that cannot be anymore satisfied by simple replication of wires. Thus, apart from network simplicity, NoC technologies must offer virtualization of the limited set of wires to maximize their efficiency.

As discussed above, Multicore requires NoC as a breakthrough technology that provides communication services through a simple, structured, scalable, and efficient on-chip network infrastructure based on on-chip routers, communication links and network interfaces that implement appropriate adaptation layers. NoC is a multidisciplinary synthesis of existing methodologies and technologies in several domains, such as SoC design, application traffic modeling, parallel and distributed computing, operating systems, computer networks, and reconfigurable interconnects.

Research efforts by the NoC community have focused on different domains, such as methodology, tools, architectures, topologies, end-to-end and network flow control, routing strategies, congestion and admission control protocols, from software perspective to novel circuit or micro-architecture solutions.

Since 2000, NoC is an extremely active research topic, both in academic and industrial domain. In fact, many innovative on-chip buses and NoC architectures have been proposed by academic and research institutes, including Bologna and Stanford, KAIST, KTH, LIP-6, MIT, UCSD, Manchester, Tampere and Technion, as well as industrial research labs, such as Philips Research Lab, ST Microelectronics, and VTT Technical Research Centre.

2.2 Popular SoC Buses

Buses have been successfully implemented in almost all complex System On-Chip designs. On-chip uses typically support either a specific set of features relevant to a narrow target market, or a specific processor. Different motivations, such as cost-performance and scalability issues, led to evolution of SoC architectures and shared buses. Nowadays, on-chip buses support advanced features, such as pipelining, split transaction, retry techniques, and standard communication sockets. Next, we briefly outline the architecture of the most widely used on-chip buses, explaining their main features.

2.2.1 The Peripheral Interconnect Bus

For SoC design, IP reuse becomes easier if standard buses are used for interconnecting application-specific components. The **Peripheral Interconnect** (PI) bus is an open on-chip bus standard, collaboratively defined almost 15 years ago by the European project Open Microprocessor Systems Initiative composed of European semiconductor companies (Advanced RISC Machines, SGS-Thomson, TEMIC-Matra MHS, Philips and Siemens). PI bus is a typical example of a synchronous, processor-independent, shared system bus which provides simple memory-mapped data transfers in commercial modular SoC design [259]. Several PI bus agents, such as master, e.g. processor cores, coprocessors, or DMA modules, and slave modules, e.g. on-chip memory and I/O interfaces, can be connected to a PI bus through appropriate signal interfaces. A bus arbiter periodically examines accumulated requests from multiple master interfaces, and grants access to a master using different arbitration mechanisms specified by the bus protocol. Notice that VHDL code for master, slave, and control is freely distributed. The PI bus does not provide support for cache coherency, broadcast, dynamic bus resizing, and unaligned data access.

2.2.2 The Manchester University Asynchronous Bus

The **Manchester University Asynchronous Bus for Low Energy** (MARBLE) is an on-chip bus with centralized arbitration and address decoding that does not rely on a global clock pulse [26]. MARBLE is used in AMULET3H microprocessor. It connects the CPU core and DMA controller to RAM, ROM, and other peripherals. MARBLE supports split transactions to asynchronous macrocells using two fully asynchronous multipoint bus channels. The first bus transfers the command from the initiator to the target, returning either accept or defer status, while the other transfers the response (with the read data or write data) from the target to the initiator.

2.2.3   The Palmchip CoreFrame

CoreFrame is a low-power high-performance on-chip interconnect based on three buses: CPU bus, PalmBus, and MBus [262]. PalmBus and MBus are parallel buses, rather than forming a hierarchy. The CPU bus can configure and access peripherals by connecting to PalmBus via a controller. It also connects to the MBus through a cache or bridge, providing high-speed access to data memory devices from the CPU core and peripheral blocks. Communication among bus subsystems is achieved through shared memory programming.

2.2.4   The Avalon Bus

Altera's parameterized, synchronous **Avalon bus** is used to connect together processors and peripherals through predefined signal types into a programmable SoC [9]. It is mainly used for FPGA SoC design based on Nios processor. Basic Avalon bus transactions are based on 8-, 16-, 32-, 64-, or 128-bits wide data transfers. This bus supports multiple bus masters, and provides different services, such as dynamic bus sizing and read/write streaming to connected Avalon peripherals. Altera's system development tool (called builder) automatically generates network fabric logic that supports each type of transfer supported by the Avalon interface.

2.2.5   The AMBA Bus

The Advanced Microcontroller Bus Architecture (AMBA) specification is a renowned on-chip bus architecture deployed in several chips. Three distinct buses are defined within the AMBA specification, the Advanced High performance Bus (AHB), the Advanced System Bus (ASB), and the Advanced Peripheral Bus (APB). Finally, AMBA AXI bus is described separately in Section 2.2.8.

The most interesting component of AMBA bus is the AHB. Within the AMBA architecture, AHB acts as the high-performance backbone system bus, connecting processors, on-chip memories and off-chip external memory interfaces with low-power macro cell functions. For example, in a hypothetical SoC, the AHB bus may sustain the external memory bandwidth required by the processor; apart from the host CPU, masters can be Direct Memory Access (DMA) devices, and there is usually a bridge to a narrower APB bus subsystem connecting lower bandwidth peripheral devices through standard interfaces (UART, USB, or HDD). AHB integrates advanced architectural features required for the design of high-performance, high clock frequency systems, such as burst transfers, split transactions, and single cycle bus master handover. AHB design is based on single clock edge operation, and non-tristate implementation.

The AHB bus protocol is designed to be used with a centralized multiplexer interconnection scheme, i.e. all bus masters drive address and control signals indicating the transfer they wish to perform. Then, the arbiter determines the master to communicate to all slaves. On the return path, a central decoder selects the appropriate signals from the slave involved in the transfer.

2.2.6 The IBM CoreConnect

In contrast to the master-centric view, the IBM CoreConnect architecture was developed to perceive bus operations as slave-centric. As a result, IBM CoreConnect implements multiple shared slave segments. The IBM Core-Connect architecture provides three buses for interconnecting cores, library macros and custom logic, namely

- the Processor Local Bus (PLB),

- the On-Chip Peripheral Bus (OPB), and

- the Device Control Register Bus (DCR).

PLB and OPB provide the primary means of data flow among macro elements. Since these buses differ in structure and control signals, individual macros interface either to PLB, or to OPB. The PLB usually interconnects high-bandwidth and low latency devices, such as processor cores, external memory interfaces and DMA controllers. IBM CoreConnect's advanced communication features include decoupled address, read and write data buses with split transaction capability, and concurrent read and write transfers yielding a maximum bus utilization of two data transfers per clock, Moreover, address pipelining reduces bus latency by overlapping an ongoing write transfer with a new write request, or an ongoing read transfer with a new read request.

Multiple masters are attached to the PLB macro via separate address, read and write data buses and a plurality of transfer signals. Slaves are attached to the PLB macro via shared, but decoupled, address, read and write data buses along with transfer status and control signals for each data bus.

IBM CoreConnect architecture can be used to interconnect macros in a SoC based on the Power PC 440. High performance, high bandwidth blocks, such as the PowerPC 440 core, PCI-X bridge and PC133/DDR133 SDRAM controller reside on the PLB, while OPB hosts lower data rate peripherals. The daisy-chained DCR bus provides a relatively low-speed data path for exchanging configuration and status information between the PowerPC 440 core and other on-chip macros.

2.2.7 The ST Microelectronics STBus

The STBus is a set of protocols, interfaces, primitives and architectures specifying an interconnection system, versatile in terms of performance, architecture and implementation [301] [302]. It is the result of the evolution of the interconnection subsystem developed for microcontrollers dedicated to consumer applications, such as set top boxes, digital TV, and digital still cameras.

The STBus protocol is organized into layers. Each operation is broken into one or more request/response pairs. Primitive operations have a single request/response pair. Depending on STBus type, compound operations may involve multiple pairs of packets. These packets are subsequently mapped into cells in the physical interface. Depending on the interface type, the amount of information transferred during the request phase may differ from that of the response phase. This asymmetry is pivotal for bandwidth allocation.

Three types of STBus protocols exist, each with a different level of implementation complexity and performance.

- Type 1 is the simplest protocol intended for peripheral register access. No pipeline is implemented, and this protocol acts as a Request/Grant protocol.

- Type 2 adds pipeline features and supports operation code for ordered transactions. It can be defined as a symmetric bus, since the number of response cells matches that of the request cells.

- Type 3 is an advanced protocol implementing split transactions for high bandwidth requirements, i.e. high performance systems. It supports out of order execution. This protocol is asymmetric, since response size generally differs from request size.

STBus architecture consists of several building blocks, such as STBus nodes in charge of switching and arbitration, size, type, and frequency converters for mismatched interfaces, retiming buffers designed to break critical path connections, and STBus register files that allow for interconnect customization, including selecting arbitration, or setting dynamic priorities of highly tunable arbiters.

2.2.8   The AMBA AXI

The Advanced eXtensible Interface (AXI) is the latest generation AMBA interface that targets high performance, high frequency SoC design [17]. With respect to previous AHB specification, AXI includes several key features, such as support for unaligned data transfers using byte strobes, burst-based transactions issued with a start address, multiple outstanding accesses, and out-of-order transaction completion. AXI defines separate address, control, and data transfer phases through separate read and write data channels. Thus, multiple outstanding transactions are supported by implementing five different channels.

- Two separate channels drive address and control signals for read and write operations,

- Two separate channels drive data from slave to master for read, and from master to slave for write operations.

- Finally, an additional write response channel allows the slave to signal completion of a write transaction to the master.

The AXI protocol provides a single flexible interface for packet transfer among master, interconnect and slave components. AXI admits different implementations, such as shared address and data bus, shared address bus and multiple data buses, or multilayer, with multiple address and data buses.

The AXI protocol supports several advanced features which represent current state-of-the-art in on-chip bus architectures. Three types of bursts are defined, namely normal memory access, wrapping cache line, and streaming data access to peripheral FIFO locations. There is system-level cache support based on different transaction attributes (bufferable, cacheable, and allocated). Atomic operations are enabled by exclusive and locked access, while different levels of privileged and secure access are provided.

AXI architecture also features an optional low power communication interface that targets two different types of peripherals. Peripherals that require a power-down sequence can have their clocks turned off only after they enter a low-power state, while peripherals that require no such sequence independently indicate when it is acceptable to switch off their clocks.

2.2.9   Wishbone

Wishbone bus architecture was originally developed by Silicore Corporation to fulfill requirements of modern SoC designs [385]. In August 2002, Open-Core, an organization that promotes open IP core development, released a public version of the Wishbone on-chip bus specifications, with the aim to spread out open core use. Wishbone is a scalable bus architecture based on simple master/slave handshake communications implemented as register read/write operations. It can be configured with an 8-, 16-, or 32-bit wide bus'to interconnect many already developed compatible IP blocks.

Wishbone provides several different interconnection topologies to the designer, such as

- direct point-to-point connection between master and slave cores,

- dataflow interconnection based on linear systolic array architectures for DSP algorithm implementation, and

- shared bus and crossbar switch interconnection, commonly used in SoCs.

Similar to other on-chip bus solutions, Wishbone also supports different types of bus transactions and arbitration schemes.

2.2.10   The MIPS SoC-It

MIPS has developed the 5×5 on-chip SoC-It switch connecting together in any combination a MIPS core, a dual port memory controller and three core interface units. SoC-It provides a high-performance solution to interconnect MIPS processor cores either to a native MIPS processor bus or multiple third-party cores, such as AMBA and PCI bridge controllers. Leveraging the multi-layer capabilities of AMBA, SoC-It offers plug-and-play usability, but thus far it is limited to PCI, AMBA AHB and a peripheral bus. The basic arbitration scheme is round-robin, but several user defined schemes may be developed. Furthermore, vectored interrupt modules have been integrated to SoC-It.

2.2.11   The SONICS Silicon Backplane

Silicon backplane is the first product of Sonics' SMART interconnects. It is an innovative, highly configurable, scalable on-chip bus that addresses SoC integration at the IP core and system architecture level.

The SiliconBackplane μNetwork interconnect is based on communicating agents that collectively manage all SOC communication, including data transfer, control, debug and test flows. Each agent is attached to an external IP core, e.g. CPU, DSP, local memory, DMA, or peripheral I/O resources. This concept decouples core functionality from inter-core communication, a crucial feature for designing complex systems. External IP cores communicate with their attached agent through μNetwork interface ports using the Open Core Protocol (OCP) which provides a standardized, configurable point-to-point socket interface. Another key point in the Silicon backplane design flow is OCP-based configuration on a per core basis. This is supported by Sonics' Fast Forward Development Environment, a socket-based tool that configures agents according to requirements of the cores and communication traffic constraints of the application.

Agent technology enables and fosters SoC development as hierarchical tiles, i.e. independent collections of functions requiring minimal assistance from the rest of the die. Through agent technology and configurability, SiliconBackplane's tile-based architecture manages to increase reuse across a SoC product family, thereby improving time-to-market.

2.2.12   The Element Interconnect Bus for the Cell Processor

The Element Interconnect Bus [68] [190] is an on-chip communication bus internal to the Cell processor which connects twelve components: the PowerPC processor, the memory controller, the eight coprocessors, and two off-chip I/O interfaces. EIB consists of an address bus and four 16 byte-wide data rings, two running in clockwise order and another two in counter-clockwise. Any bus requestor can use any of the four rings to send or receive data. Moreover, each ring can potentially allow up to three concurrent data transfers as long as their paths do not overlap. To avoid stalling of read requests, data ring arbitration assigns priority to the memory controller, while remaining requests are treated equally with a round-robin priority. The maximum network bandwidth is limited by the snooping protocol (and not bus concurrency) to 204.8 GBytes/sec for a 3.2 GHz system clock. However, experimental tests with mainly scientific loads carried out by IBM indicate a sustained effective data bandwidth between 78GB/s to 197GB/s (96% of peak). Sustained data bandwidth is lower than the peak, since several factors affect the efficiency of the data arbiter, such as many-to-one access to memory or to one of eight local stores, relative location of the destination and source nodes (paths cross), and possible interference of new transfers with ongoing ones.

2.3 Existing NoC Architectures

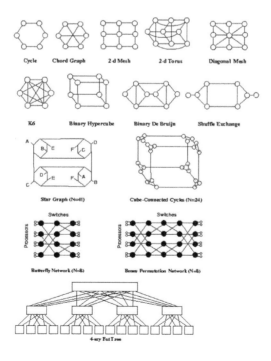

FIGURE 2.8: Different NoC topologies

Several currently proposed NoC interconnects target homogeneous Multicore processors. Similar to parallel interconnects, homogeneous Multicore processors focus on regular topologies, usually 2-d mesh [251]; see also the NoC instances discussed later in this Section. Other innovative NoC architectures are based on common topologies, as illustrated in Figure 2.8. More complex network topologies, such as expander graphs, e.g. multi-butterflies and multi-Bénes networks, are very powerful in terms of global communication capabilities, especially for large number of nodes [208]. However, they are not practical for current NoC network sizes.

Moreover, there are several examples of NoC communication for embedded consumer devices based on Multicore SoC architectures. In these architectures, component heterogeneity has been addressed by application-specific topologies.

In this Section, we provide additional insights on interesting NoC architecture instances and design approaches, such as the LIP6 SPIN, MIT Raw, Philips Æthereal, Silistix Chainworks, University of Bologna and Stanford's xpipes, and Arteris NoC.

The **Scalable Programmable Integrated Network** (SPIN), proposed by the University of Pierre and Marie Curie - LIP6 laboratory, constitutes one of the first concrete implementations of switching networks on silicon chips, thus proving the feasibility of NoC as an interconnection medium for embedded systems. The point-to-point SPIN architecture is based on a 4-ary fat-tree topology implemented with 8×8 routers, adaptive wormhole routing and input queuing [14] [15] [132]. A fat tree is a tree structure with routers on the nodes and terminals on the leaves, except that every node has replicated fathers. The SPIN network size grows with a $1.25NlogN$ factor to the number of leaves N (or IP blocks). SPIN links are point-to-point, full-duplex, 36-bit wide in each direction and use credit-based flow control. There is no limit on packet size, since each packet consists of a sequence of 4-byte flits, i.e. first flit, data flits and end of packet flit. The first flit (header) has a 1-byte address and remaining bits for special services or routing options. Although IP nodes at level 0 outnumber routers at level 1 by a ratio of (4:1), for all other levels we have the same number of routers. Moreover, for every router, the number of parents is the same as the number of children in the tree. These properties imply that the network has a high bisection throughput. In addition, by exploiting routing path adaptivity, routers are able to utilize redundant network paths, and thereby reduce network contention, providing tradeoffs between performance and power consumption.

SPIN uses small (4 word) input buffers designed to hide link latency and control logic delay. It also contains two (18 word) output buffers to handle output contention, thus internally it implements a 10×10 partial crossbar. It is a scalable network for transporting packet data, but uses a bus network for control. Since it is a best-effort network optimized for average performance using flow control, there is commitment for packet delivery, but latency bounds are only statistically given.

Its network interface consists of two wrappers (VCI/SPIN and SPIN/VCI) compliant to Virtual Component Interface (VCI) specifications for interfacing the SPIN network with external IP [373]. In 2003, a 32-port SPIN network was implemented in a 0.13u CMOS process with total area 4.6 mm^2 and peak bandwidth ~100Gbits/s.

In the wormhole-routed **Butterfly Fat Tree** NoC topology (BFT), the layout is a fat tree [263]. Each tree node is represented with a pair of coordinates: its level in the tree (starting from zero at the leaves) and its position, e.g. in right to left ordering. Tree leaves correspond to IP or processor cores. Each switch at layer one is connected to four such blocks. Intermediate routers (at layer two and above) are allocated two parent ports, and four child ports, or connections, i.e. BFT is a 4-ary fat tree. Notice that the number of routers at each level decreases by a factor of two. Thus, BFT provides an overall cores to routers ratio of 2:1, compared to 4:1 for mesh. However, this architecture enables simple traffic aggregation to/from a particular set of cores and regular structuring of the switches in the layout, simplifying design. In fact, BFT trades throughput for reducing area overhead and power efficiency, more than the SPIN. For a more precise definition of general fat tree topologies refer to Section 3.5.

The **MIT Raw** microprocessor network is a research prototype that attempts to reduce the emerging large wire delays by implementing a simple, highly parallel VLSI architecture with a scalable, configurable instruction set architecture [352] [284]. Raw software fully exposes low-level details to the gate, wire and pin resources of the chip, enabling the programmer (or compiler) to determine and implement the best allocation of resources, including scheduling, communication, computation, and synchronization among PEs, for each possible application.

The MIT Raw design divides silicon area into 16 identical, programmable tiles. Each tile consists of an eight-stage, in order, single issue, MIPS processor, a four-stage, pipelined floating point unit; a 32-Kbyte data cache and 96 Kbytes of software-managed instruction cache. Each tile connects to north, east, south or west neighbor tiles using four full duplex 32 bit-wide networks, two static and two dynamic. The static router (routes specified at compile time) is a five stage pipeline that controls two physical networks used for point-to-point scalar (operand) transport among tiles. The dynamic routers control two dynamic networks for remaining traffic, e.g. memory, interrupt, I/O, and message passing. This implementation reduces the longest wire, ensuring scalability. Packets consist of a single header that specifies the destination tile, a user field and packet length. Raw implements fine-grain communication between replicated processing elements with local memory. Thus, it is able to exploit parallelism in data parallel applications, such as multimedia processing.

The **Embedded Chip-Level Integrated Parallel SupErcomputer** (Eclipse) is a scalable high-performance computing architecture for NoC by VTT and various Universities [113]. Processing elements are homogeneous, multithreaded, with dedicated instruction memory, and highly interleaved (cacheless) memory modules. The network interconnect is a high capacity, 2-d sparse-mesh with routers in the middle and processors only in the mesh peripheral. This structure provides enough bandwidth for heavy random communication and exploits locality and avoids memory hotspots (and partly network congestion) through randomized hashing of memory words around a module's memory banks. The programming model is a simple lock-step-synchronous EREW (Exclusive Read Exclusive Write) PRAM model.

The **Chip-Level Integration of Communicating Heterogeneous Elements** NoC architecture (called Cliche) is based on a simple 2-d mesh topology with a store-and-forward communication protocol, i.e. first the complete packet is stored locally, and then, if there is enough space, it is forwarded to the next node [195]. Each router is connected through input/output channels to one external IP or processor/DSP resource and three or four other neighbor routers, depending on its precise location in the layout. Connection to the external resource may require network adapters. The network communication load can be reduced by implementing specific packet admission mechanisms. Moreover, for moderate network traffic loads, the probability of a packet drop can be almost eliminated by implementing buffers that can hold a large number of messages (e.g. practically 8).

The **Nostrum chip** developed by researchers at KTH University and the research group working on Eclipse is an application-specific, low power NoC platform based on Cliche. Nostrum interconnects embedded systems using a traditional 2-d mesh consisting of heterogeneous processor and 5×5 router tiles. Nostrum implements adaptive deflection routing without buffers or routing tables [234] [235], but with congestion detection. The basic services are best effort, while a guaranteed latency is achieved through virtual circuits, using an explicit time division multiplexing mechanism called temporally disjoint networks (TDN). The resources are heterogeneous and their computing and storage capacity is designed according to application-domain characteristics. The Nostrum design flow is a combination of platform- and distributed system-based design, emphasizing reuse, separation of infrastructure and application functionality, error correction, and voltage swing for power savings. In 65nm CMOS technology, they foresee a $22mm \times 22mm$ chip size with $2mm \times 2mm$ resources and $100\mu m \times 100\mu m$ switch.

The **Æthereal NoC** developed at Philips Research Laboratories is a full-fledged on-chip interconnect consisting of generic routers and network interfaces [288] [125] [126] [124] [279]. The routers use input queuing, deterministic source-based (destination tag) wormhole routing and link-level flow control. To provide time-related guarantees, such as throughput guarantees (on a finite time scale) or latency bounds, interference of other traffic must be limited and characterized. The Æthereal NoC facilitates software programming by offering a strong end-to-end QoS paradigm that provides high-level services, such as transaction ordering or throughput and latency guarantees; guaranteed services are used to express composition and robustness in critical real-time applications. The QoS protocol defines traffic classes for throughput/latency guaranteed (GT) and best effort service (BE). While GT flits use a connection-oriented, contention-free time-division-multiplexed circuit-switching based on slot tables and appropriate packet headers, BE flits are scheduled to remaining output ports using conventional wormhole routing, input or output queuing, and round-robin arbitration.

Since all network routers implement a common notion of time in a slot counter, link contention for GT packets is completely avoided by controlling the time they enter the network using a local slot allocation table. Notice that slot allocation can be performed statically during initialization. In recent versions, in order to save area, slot tables have been removed, while this information is provided in the GT packet header. BE traffic is scheduled to non-reserved, and reserved but unused slots with a non optimal algorithm based on parallel iterative matching. The logically separated guaranteed (GT) and best-effort (BE) routers are combined to share link and data path resources. In order to reduce area costs, input queuing is implemented using custom-made hardware fifos.

The Æthereal network interface converts the OSI network layer of the routers to transport layer services for the connected IP. All end-to-end connection properties are implemented by network interfaces, i.e. reordering, transaction completion and flow control. IPs negotiate with network interfaces to obtain connections by reserving resources, such as network interface buffers, credit counters and slots in router tables. Æthereal supports narrowcast, multicast or simple connections and shared memory-like transactions, such as read, write, acknowledged write, test and set or flush.

An Æthereal router has been prototyped with $0.26mm^2$ area in CMOS12 and offers an 80GByte/sec peak throughput. The Æthereal network interface has an area of $0.25mm^2$ in 0.13 μm, running at 500Mhz.

Furthermore, an automated design flow for instantiation of application-specific ÆTHEREAL has been described [270]. The flow uses XML to input various parameters such as traffic characteristics, GT and BE requirement, and topology.

The **PUCRS Hermes** infrastructure generates wormhole routed NoCs based on basic network components, such as routers and buffers [238]. Hermes employs different topologies, while adjusting flit and buffer size and routing algorithms. Hermes implements three layers of the OSI reference model: physical wiring interface, an explicit handshake data link protocol for transferring data reliably between routers and a network layer for packet switching. It also supports OCP, ensuring enhanced reusability of the infrastructure and connectivity to available, compliant IP cores. The main component is the HERMES router which aims at a 2-d mesh topology. It contains control logic and a set of up to 5 bidirectional ports, i.e. East, West, North, South connections to 4 neighbor routers and local connection to an IP core. Hermes can use simple XY routing with input queue buffers to enable practical small area implementation. In addition, dynamic arbitration resolves conflicts when multiple packets arriving simultaneously at the router require the same output port.

The Technical University of Denmark's **Mango** is a message-passing asynchronous NoC implementing coarse-grain GALS with delay insensitive, encoded links [39] [40]. It also provides connection-less best-effort routing, as well as guaranteed services using virtual channels; these can be implemented using separate physical buffers and a smart scheduling scheme called asynchronous latency guarantees [41]. Notice that for this scheme, latency guarantees are not inversely dependent on bandwidth guarantees, as is the case of TDM-based scheduling. In addition, this makes global timing robust, since no timing assumptions are necessary between routers. Mango interfaces the asynchronous network to clocked OCP-based standard socket through network adapters designed using primitive routing services of the clockless network.

The **Intel Teraflop NoC** design is based on an 80 processor multicore system, configured as a 10×8 2-d mesh network. Each core, consisting of a dual floating point unit, and local instruction and data memory, is connected to a fast, compact, message-passing 5-port router through mesochronous links that allow for energy-efficient processor and router circuits (through clock gating). More than 100 GB/s node bandwidth results in a total 2.6 terabits maximum network bisection bandwidth at only 62 Watts [367]. The Teraflops research chip could enable future applications in instant video communications, realistic PC-based game entertainment, multimedia data mining, and handheld real-time speech recognition.

ANOC is an asynchronous NoC for telecommunication applications proposed by CEA/LETI within a European Medea+ project [114]. The NoC is integrated in the research-oriented multicore SoC architecture called FAUST which consists of an ARM946 core, embedded memories, smart DMA engines, numerous highly programmable hardware blocks and reconfigurable data-paths engines. NoC implements a 2-d mesh with adaptive non-minimal path wormhole routing based on the turn model. In addition, asynchronous communication is based on a quasi-delay-insensitive logic, while NoC functional units are implemented with independent clock domains using standard synchronous design methodologies. Although ANOC uses virtual channels, there is no support for QoS.

NEC Labs [117] [118] have proposed a heterogeneous tile-based architecture configured as a 2-d mesh or torus topology, customized during the application mapping phase according to constraints, such as bandwidth. Each tile consists of a processing unit, memory system, dedicated hardware components, and one or more internal bus subsystems. It also provides a wrapper for connecting to four other neighbor tiles through input and output channels; this wrapper essentially acts as a network interface by providing packet routing and buffering facilities to handle congestion. Dedicated receiver and sender units act as adaptation layers (interfaces) between internal bus and tile wrapper, allowing internal bus modifications without modifying the tile wrapper.

The **KAIST BONE** (basic on-chip network) research project [206] is based on a power-efficient, packet-switched, hierarchical star topology, essentially an $\sqrt{N}$-ary tree. BONE focuses on research on NoC protocols and architectures and targets system-level and HDL simulation and analysis, and circuit-level design, implementation and functional validation of a heterogeneous NoC. The fabricated chip emulates a large scale, low power SoC for an embedded system containing several heterogeneous IPs, such as multiple RISC cores, SRAM memories, an off-chip gateway, a reconfigurable logic array, and peripherals operating at different clock frequencies.

QNoC [133][43] is an interesting wormhole-routed irregular 2-d mesh network which guarantees QoS at four different service levels. The first level minimizes system response time for critical OS interrupt messages and control signals, e.g. for memory access. The second level provides hard and soft real-time guarantees for latency, jitter (temporal spread of latencies) and bandwidth in streaming audio and video applications. The third level deals with read/write bus semantics in register/memory access, and finally, the last level performs long data transfers, such as DMA.

The QNoC router allocates small independent virtual channel buffers for each service level and implements round-robin scheduling within each service level and preemptive priority scheduling among all service levels, with bounded traffic for high priority to avoid starvation. Thus, during QNoC arbitration, each output port schedules flit transmission based on buffer availability at the next router (stored in separate tables for each service level at each output port), priority (service level), and a round-robin order of the input ports awaiting packet transmission within the same service level. This implies that a particular flit may be transmitted on an output port if there is buffer space available at the next router and no pending packet with a higher priority for that particular output port.

QNoC network implements deadlock-free shortest path XY routing and hop-to-hop credit-based control to ensure lossless flit transmission. Another important QNoC feature is the ability to construct an asymmetric network with variable speed links. Different network links, connecting routers to other routers or modules, may be set to different bandwidths during the design process to meet QoS requirements. Moreover, QNoC multiplexes packets on the active outgoing virtual channels over the physical link in round-robin manner for increased performance.

The **CHAIN NoC** [27] uses ad-hoc topology implementing GALS with asynchronous links based on delay insensitive 1-of-4 encoding to interconnect IP modules efficiently, providing best-effort services while considering also prioritization in asynchronous networks [110] [109]. In fact, Chain asynchronous design is related to MARBLE bus design (see Section 2.2). Chain provides a flexible, clock-independent solution, increasing bandwidth, reducing power consumption, and resolving timing closure problems in deep submicron technology. A router implementation for CHAIN provides differentiated services with soft deadlines by prioritizing VCs [110]. The Silistix Chainworks EDA

library toolset supports development of custom, self-timed NoC using industry accepted platform-based design and tool flows. Chainworks consists of a
library and two tools that plug into traditional ASIC/SOC design flow.

The **Hibi NoC** [295] is a communication network aiming at maximum
efficiency and minimum energy per transmitted bit. It includes QoS based
on distributed TDMA and supports hierarchical topologies with several clock
domains. It is used to integrate coarse grain components that have the size
of thousands of gates.

The **Sonics' SMART SonicsMX Interconnect** provides a full suite of
efficient, energy-aware communication services for heterogeneous Multicore
based on advanced agent technology [327].

SonicsMX is an innovative pipelined, multithreaded, non-blocking interconnect that uses traditional synchronous design methodology. It is based on
highly optimized and flexible (full or partial) crossbar structures, shared bus,
or hybrid topologies with split transactions. SonicsMX targets low-power,
cost-effective SoC design for advanced wireless communications, multimedia-
enabled PDA, and digital consumer electronics, such as HDTV, digital camera, multifunction peripheral, SetTop box, and video games. SonicsMX provides seamless connectivity to Open Core Protocol (OCP) versions 1.0 and
2.0, and ARM's AMBA AHB or AXI cores. Automatically generated agents
within SonicsMX cleanly decouple functionality of each SoC core from the
NoC interconnect. SonicsMX provides multi-level quality-of-service ensuring
through priority arbitration guaranteed bandwidth or latency paths between
specific (initiator, target) pairs, configurable (instead of fixed) power management changing voltage levels and clock rates of the SoC cores, as well as
security protection and error handling.

SonicsLX is a light-weight, upward compatible implementation of a strict
subset of SonicsMX for cost-effective mid-range SoC design. It provides simplicity, low cost, and small chip area and it can be seamlessly upgraded to
SonicsMX.

The SonicsStudio tool suite is an integrated SoC development environment
for SonicsMX or SonicsLX solutions. It allows automated and accurate configuration (RTL or SystemC models), simulation test bench, display and analysis, data analysis and exploration, and performance verification. It seamlessly
connects together parametric IP, including control processors from IP vendors, the Sonics DRAM memory scheduler (called MemMax), or peripheral
interconnects (Sonics SC320). In addition, SonicsMX provides SystemC (and
equivalent RTL) models for system modeling, design, verification, and synthesis.

The **folded torus NoC** layout provides extra wiring, reducing the hop
count for any transmitted packet. However, reducing misrouted or dropped
packets and avoiding deadlock (through virtual channels) require a larger
buffer size. In fact, evaluating tradeoffs between buffer requirements and
performance for different flow-control schemes is fundamental for designing
an efficient torus NoC [89].

The UFRGS **System-on-Chip Interconnection Network**[395] (SOCIN) is a scalable, low cost NoC based on 2-d torus topology based on the Router Architecture for System-on-Chip (RASOC) parametric routers [394]. The routers have five communication ports: east, west, north, south and the local port that is used to connect the router to a specific processor through either directly, or through a custom bus. Routers implement wormhole routing and credit-based flow control. Experiments have used 16-bit or 32-bit wide channels (similar to flit size), and packets of unlimited length. SOCIN uses credit-based flow-control and XY routing, a deadlock-free, deterministic source-based approach, in which a packet is firstly routed on the X direction and then on the Y direction. Routers include 4-flit queues at each input port and round robin arbitration.

The **Adaptive System–on-Chip** (aSOC) is an on-chip communication architecture designed to promote scalability and flexibility in system-on-chip design [211]. aSoC implements a power-efficient statically scheduled 2-d mesh topology and includes a companion compiler CaSoC. ASoC provides power-aware features of modern cores, such as automatic runtime selection of voltage and frequency of each core by monitoring interconnect utilization. A simple clock division scheme makes it possible for each core to select and switch between 8 different frequencies. In addition, a voltage scaling procedure allows the individual core supply voltages to switch between 4 different values. This technique is able to reduce power up to 90%, as shown with video encoding systems in portable digital signal processing applications.

The University of Linkoeping **SoCBUS** [381] is a 2-d mesh network implementing a packet-switched circuit to set up routes through the network, i.e. a packet is switched through the network locking the circuit as it moves. Although this leads to deterministic behavior, it restricts communication flexibility.

×**pipes and the NetChip compiler**, developed by University of Bologna and Stanford University, provide an automated platform-based NoC design flow based on parametric network building blocks for customized, application-specific NoC architectures [246]. ×pipes library provides switches that support reliable communication for arbitrary link pipeline depths (links can be pipelined with a flexible number of stages), and an OCP-compliant network interface that connects to/from IP cores. ×pipes supports regular and heterogeneous architectures with source-based, wormhole routing. Based on the eventual SoC architecture floorplan consisting of switches, links, network interfaces and IP blocks, the ×pipes compiler automatically extracts synthesizable SystemC cycle- and signal-accurate executable specifications for all network components. Input to the ×pipes compiler is provided either through a user-specified topology file, or using the SUNMAP tool. This tool automatically maps IP cores onto NoC topologies selected from a library, considering different NoC routing strategies, such as dimension ordered and shortest path. By utilizing floorplan information, SUNMAP can minimize area or power dissipation requirements, and maximize performance characteristics.

The **Arteris Danube interconnect library** consists of a set of configurable IP blocks (mainly switches and network interfaces) that manage on-chip communication, thus enabling the design of customized application specific topologies. The layering approach is implemented by splitting the interconnect functionality into decoupled layers; switches deal with transport-layer packets, while network interface units manage the transaction level. NoC communication is based on a proprietary packet-based protocol, while IPs are connected through network interface units that support standard socket interfaces AMBA AHB, AMBA AXI, and OCP. Arteris interconnect transports packets over point-to-point connections with 32 and 64 bits synchronous or mesochronous GALS links. Non-intrusive runtime application debug and service units provide NoC monitoring and control to embedded software, while a special DRAM memory scheduler improves external memory efficiency. The Danube library is part of an integrated flow for efficient design space exploration and for generating an interconnect instance, as a SystemC, or VHDL model, aided by the Arteris NoCexplorer tool and the NoCcompiler configuration and assembly environment.

ST Microelectronics' Versatile Network on Chip [363] (VSTNoC) is a packetized, pipelined multi-stage crossbar. Although VSTNoC is largely based on the existing STBus design approach, it is a first step towards the evolution to a complete NoC solution. In this respect, VSTNoC provides a versatile family of topologies and a packet format (header + payload) based on communication layering. In addition, the socket protocol for external IP is based on a new component called a network interface unit. The highest level transmission entity in VSTNoC arbitration is a transaction which consists of a collection of packets with header information fields (e.g. routing, QoS, packet transfer size, write or read opcode) and payload. VSTNoC reduces wire density using header/payload multiplexing techniques, thus simplifying physical design issues, such as wire congestion, parasitic effect, and mapping onto FPGA. In addition, operating frequency increases due to a pipelined node architecture, and chip area decreases due to wire reduction and removal of dependencies between send and receive paths.

Based on Real World Technologies analysts [286], Intel's The **Common System Interface** (CSI) is a new (expected in 2008, renamed as "Quickpath") point-to-point interconnect for multi-socket server systems that support an on-chip memory controller. It is a layered network fabric based on five distinct layers: physical, link, routing, transport and protocol layer which provides cache coherency protocols based on MESI. Connected cores, such as microprocessors, coprocessors, FPGAs, and chipsets must provide a CSI port. Compared to the existing Intel parallel P4 bus, CSI uses vastly fewer pins running at much higher data rates. Thus, due to reduced memory and remote access latency, Intel's microprocessors will need less cache, freeing area for additional cores, or more economical die sizes. CSI supports advanced power saving techniques and hot-plugging, i.e. swapping out of a processor, memory or I/O hub without bringing down an entire node. There are five initial CSI implementations in Intel's 65nm and 45nm high performance CMOS processes that provide 12-16 GBytes/s of bandwidth in each link direction.

Chapter 3

NoC Topology

3.1 On-Chip Network Topology

Market, application and technology trends impose new challenges for Multicore design and on-chip interconnects. As explained in Chapter 2, a radical shift towards distributed packet-switched NoCs is mandatory.

A NoC is characterized by its topology which describes the physical interconnection structure of the network graph, the routing algorithm which determines the set of paths that packets may follow, the switching strategy which defines the mode of operation, e.g. circuit switching vs. packet switching, and the flow control mechanism which governs network resource allocation. In particular, NoC topology defines a communication infrastructure based on basic switching elements, usually routers and links that can capture today's Multicore requirements. In fact, the need to interconnect NoC resources in such a way that will ensure optimal communication performance is paramount and topology choice has a significant impact on Multicore cost-performance.

Notice that the routing algorithm is intrinsically related to the network topology, since it describes the rules that regulate packet transfer through the network topology. Complex routing algorithms previously deployed in computer networks and parallel interconnects are generally not suitable within the NoC domain. NoC point-to-point routing must be based on simple and efficient routing algorithms without routing tables or complex arbitration protocols, targeting a small area and high frequency implementation. Routing schemes define precisely how network packets move through the communication infrastructure. Moreover, smart routing decisions reduce traffic congestion and improve reliability and fault tolerance possibly by resorting to locality and by efficiently mapping in a scalable way common communication patterns in different SoC applications. Such compact routing schemes include source-based (originally called destination tag), function-based (e.g. mono- or multi-level interval), prefix-based, and hierarchical routing (e.g. dimension-order) [129]. Adaptive and probabilistic routing techniques usually cause data reordering hazards and should be avoided in NoC context.

In NoC literature, proposed architectures can be classified into topology-dependent, i.e. based on a fixed, well-defined topology, or topology-independent, i.e. the building blocks can be composed into customized interconnect structures depending on the specific application. Thus, in general, multicore processors deal with a regular network topology, while Multicore SoCs tend to use topology-independent structures.

A topology-dependent NoC has a considerable degree of flexibility, e.g. in arbitration, routing, flow control, queue size, or QoS algorithm, which can be tuned to target specific application requirements. In addition, a specific regular topology is usually considered as a member of a quite large family of regular structures which maintain similar properties, e.g. identical routing logic and similar deadlock avoidance. Moreover, a regular topology may not necessarily correspond to a regular layout implementation, but to regularity in the functional graph description.

A topology-independent NoC is versatile in its structure, but this approach has severe drawbacks that complicate its customization, since a considerable effort is required to select the optimal topology given the large design and verification space. Building blocks are required to be designed with a high degree of flexibility required by the custom topology, therefore limiting cost optimization opportunities or imposing complex architectural choices in order to support the required utilization freedom, e.g. address decoding tables rather than simple on-the-fly routing functions. Moreover, interconnect design complexity and flexibility increase the risk of routing-dependent deadlock, imposing careful investigation of each interconnect instance.

The structured nature of the NoC is a crucial point in designing a scalable platform. However, in this respect, optimum interconnect design is neither time-effective nor strategic as much as other architectural aspects. These aspects include reuse, protocol interoperability, packet structure, link performance parameters, and router and network interface design characteristics, e.g. flow and congestion control, number of virtual channels, buffer size, packet arbitration, link scheduling and QoS.

The Multicore SoC nature leads to application-specific interconnect topologies which target efficient resource utilization and low power consumption. Although automatic or semi-automatic topology generation methodology aims to improve productivity [150], it has not been used effectively in real products, since the hypothesis by which the topology is generated cannot capture real application requirements. Due to deep submicron issues in 45nm technology and beyond, regular network structures with attractive properties appear to be the right solution for Multicore. In fact, specific customization that reflects heterogeneity can be achieved through two key concepts which can be considered on top of regular structures: hierarchy and aggregation. Hierarchy of regular topologies and aggregation of traffic at the boundary of the network can be applied to match application requirements and reduce network size and cost [183]. Aggregation combines traffic from different nodes into a single network interface, thus reducing average hop count and improving latency. Robust and structured application-specific interconnects must be built on top of regular topologies, by considering area and cost constraints. Moreover, depending on features, it is possible to achieve a simplification of components through elimination of logic and channels not needed by a specific application traffic.

Before focusing on topology-dependent NoC design, it is important to distinguish between direct and indirect on-chip networks.

FIGURE 3.1:   Examples of regular network topologies

In direct packet-switched on-chip networks, each network router is directly connected to a usually small number of other network routers through (multiple) pairs of unidirectional links in opposite directions. Moreover, each router is connected through a network interface to one or more input and/or output devices, i.e. processors, memory, and external IP. Figure 3.1 shows examples of direct topologies used in different NoC solutions.

In the indirect networks the ratio of router nodes to device nodes is greater than 1 to 1. In fact, some router nodes simply connect other routers. Examples of indirect network topologies are the crossbar and multistage network (see Section 3.2).

The body of knowledge on interconnection networks already available from computer networks, parallel and distributed systems and telecommunication is a virtually infinite source of results waiting to be applied to the NoC domain. This is not simple, since constraints and requirements imposed by low-cost silicon implementation and Multicore application traffic are significant and fundamentally different than parallel interconnects and computer or telecom networks. Thus, it is very unlikely that interesting, asymptotically optimal, but complex topologies and routing methods will find applicability in Multicore design.

Current parallel systems generally use commodity microprocessors, thus memory, communication functions and I/O require separate chips. In contrast to parallel systems that extend over many printed circuit boards, with each board containing only one or more processors according to packaging, NoC should be used with multiple, usually heterogeneous and specialized computing resources on a single chip.

In the last Section of this Chapter, a panorama of NoC topologies is provided, and the Spidergon topology is introduced and evaluated from a theoretical standpoint.

3.1.1   Theoretical Metrics for NoC Topologies

Alike other interconnection networks, NoCs are characterized using several graph theoretical and combinatorial metrics. Topology comparison is usually based on theoretical cost and performance metrics, such as number of nodes, network degree, diameter, average distance, bisection width, number of edges, extendibility, symmetry, routing strategy, as well as embedding properties for common communication patterns. The most important metrics for NoC design are described below.

Network size refers to the number of device nodes connected to the network. NoC domain requires instantiating networks of different sizes with fine network size granularity.

Network cost is determined usually by the number of routers, number of cross points, communication links, wire length, wire density, and planar or multilayer VLSI layout area complexity models.

Extendibility is a desirable property that indicates if it is possible to create larger networks by simply adding new nodes without changing the topology. The metric associated to this property is the minimum granularity for scaling a topology.

Node degree refers to the number of edges connected to a router node. The network degree is defined as the maximum node degree for all nodes in the network. If the node degree is small, then the number of links and communication complexity of the router, including wire congestion in the physical layout, are likely to be small. However, a small degree implies less connectivity and larger distances among nodes. Moreover, if the node degree is constant, independent of the network size, then the network is more modular, since a single universal router is used, independent of network size.

Edge bisection width refers to network wire density. It is defined as the minimum number of edges that must be cut to separate the network into two equal halves (within one node). More specifically, a network cut $C(N_1, N_2)$ is a set of channels that partition all N network nodes into two disjoint sets N_1 and N_2. If the disjoint sets have the same cardinality (within one node), i.e. $|N_2| < |N_1| < |N_2| + 1$, then the cut is called bisection. This is an important metric, since the rate at which information crosses the network bisection (called bisection bandwidth) is the product of the bisection width (number of links), the number of wires at each link (link width) and the data transfer rate of link (link bandwidth). For continuous routing with independent, uniform random traffic, a large bisection increases VLSI complexity, but provides nice distribution among different paths, thus reducing communication bottlenecks, i.e. latency and saturation rate.

Edge bisection is notoriously difficult to compute. For example, the bisection of the chordal ring in Figure 3.2 is small, although it appears otherwise. The maximum bisection width ($N^2/4$) is exhibited by the complete graph.

Network diameter refers to the maximum number of edges along a shortest path route connecting any two network nodes, i.e. the hop count of a short-

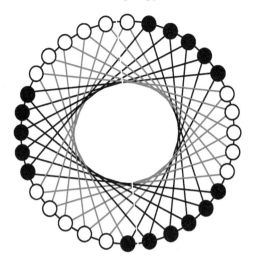

FIGURE 3.2:   Computing the edge bisection of a graph

est path. The shortest path is the minimum number of edges which must be traversed in order to connect two network nodes. For all switching strategies, a small network diameter improves network contention (in buffers and edges) and reduces propagation delay in point-to-point, since fewer routers, buffers and edges must be traversed when communicating between a pair of nodes.

Average distance is the average number of hops along all shortest-path routes in the network. A small average distance improves saturation rate and average latency for continuous packet routing with independent, uniform random traffic.

Network symmetry is important for resolving complex VLSI design issues, implementing efficient point-to-point routing and intensive parallel communication, and providing efficient emulations of topologies or embedding of parallel algorithms.

- A graph is vertex-symmetric if an automorphism exists that maps any node a into another node b, i.e. the topology looks identical from any network node.

- A graph is edge-symmetric if an automorphism exists that maps any channel a into another channel b, i.e. the topology looks identical from any network link. Notice that edge symmetry also implies vertex symmetry. In general, an edge (and vertex) symmetric topology provides the advantage of simple routing and scheduling decisions, since all edges (or nodes) have global knowledge of the network. Thus, the router hardware implementation is expected to be simple and routing decisions fast.

Network emulation and **graph embedding** of common communication patterns are also fundamental. For example, existence of a Hamiltonian path or cycle (covering all network nodes through a network path) is fundamental for implementing efficient one-to-one or many-to-many communication and parallel sorting problems. The Hamiltonian path is also used to number processing elements successively.

Network partitioning (or hierarchical recursion) refers to decomposing a higher-dimensional graph into independent lower-dimensional subgraphs of the same topology. This property holds for all topologies which can be defined as a Cartesian cross product, such as tori. This property relates to improved fault tolerance and efficient programmability, since it enables multitasking and recursive decomposition of large-scale applications.

Connectivity must be taken into account in any practical comparison of interconnection networks. Connectivity (or robustness) metrics measure network topology resiliency, i.e. its ability to continue operation despite disabled components. High connectivity implies alternative routes for delivering packets in case of router or link faults, or communication congestion. We distinguish between two types of connectivity.

- Node connectivity is the minimum number of nodes that must fail in order to partition the network into two (or more) disjoint networks.

- Link connectivity is the minimum number of links that must fail in order to split the network into two (or more) disjoint networks.

3.2   Multistage Interconnection Networks

In the 1950s, telephone switching applications led to cost-effective design and implementation of multistage interconnection networks with multiple layers of switches (MINs). MINs initially implemented circuit switching to eliminate delays after a physical connection was established, thus providing guaranteed service for time-critical applications, such as voice transfer.

Circuit switching establishes a dedicated circuit between terminal nodes in a network composed of switching elements, connected together with full duplex point-to-point links. Thus, once a connection is established, capacity is guaranteed until the call is disconnected. Notice that capacity is wasted on preset connections, if they are not used continuously. Switching elements may operate in time, by employing arbitration protocols to allocate time slices on time-division multiplexed links, or space, by employing buffers to establish separate physical connections between terminal nodes.

While switching elements are usually small crossbars, specialized networks, such as generalized connectors realizing all possible input-output (many-to-

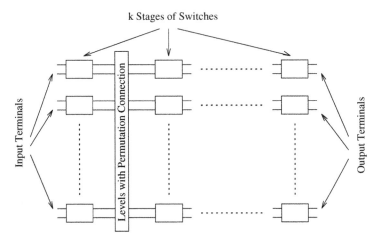

FIGURE 3.3: A generic k-stage multistage interconnection network

many) relations [72] [256] [355], are usually based on broadcasting switches, concentrators, hyperconcentrators, or superconcentrators [366] [274].

Since the 1970s, MINs have been widely used to design and implement large scale multiprocessor architectures, where usually processors communicate to each other by reading and writing data on distributed interleaved shared memories. In this case, MINs are implemented as a packet-switched network, since circuit switching is not reactive to dynamically changing bandwidth requirements, e.g. bursts in an MPEG stream. Unsuccessfully routed messages were buffered, misrouted, or dropped and retransmitted using an acknowledgment protocol.

Next, we provide a classification of circuit-switched MINs based on I/O connection assignments that can be realized simultaneously.

Blocking MINs cannot realize all permutations, i.e. no edge-disjoint paths can be drawn simultaneously for an arbitrary I/O permutation. Many blocking networks are essentially identical, even though they may be drawn differently. Two MINs are topologically equivalent if their underlying graphs are isomorphic. In wide-sense topological equivalence one MIN is obtained from the other by relabeling (redrawing) switches and renumbering I/O terminals, while in strict-sense (or terminal) topological equivalence, the numbering of I/O terminals is preserved. Two MINs are functionally equivalent if they realize the same set of connection assignments at the inputs and outputs, i.e. permutations for a permutation network. Notice that MIN topological equivalence implies functional equivalence [265] [388].

Nonblocking and rearrangeable MINs provide multipaths between every I/O pair (e.g. processor to processor or processor to memory) and can realize all I/O permutations at the same time. In nonblocking MINs a request to establish a path from a source to a destination node, which may arrive asynchronously with respect to other requests, can be satisfied without perturbing existing connection paths in the network. Nonblocking networks are strict-sense nonblocking when an I/O path is always possible irrespective of already established connections, and wide-sense nonblocking if a routing protocol is required for realizing new I/O paths to avoid interference with established paths. A rearrangeable MIN can connect any input to any free output, but with rearrangement of existing I/O paths.

3.2.1   Blocking MINs

The family of banyan networks includes all minimal, full-access MINs (all outputs are reachable from any input) with a unique path for each input-output pair [123]. Multistage banyan networks constructed from a number of fixed size $n \times m$ switches are called regular. Otherwise, they are irregular. Rectangular banyan MINs are regular banyans with $n = m$. Rectangular banyans built recursively are called SW-banyan networks. An SW-banyan is a Delta network if it has a certain level of distributivity and a unique routing path descriptor, determined by destination node only. More precisely, a Delta network consists of either

- a single switch (e.g., an $n \times n$ crossbar), or

- one stage of $n \times n$ switches, followed by one stage of n-disjoint recursively-defined Delta networks: Δ_1, Δ_2, ..., Δ_n; The buddy property defines the necessary connection requirements: input nodes in the network are connected to the first stage switches, and output j of each switch in the first stage is connected to an input node of the Δ_j Delta network [267]. The property leads to simple source-based (called destination tag) routing.

A Delta MIN which after exchanging its input and output terminals remains a Delta network is called a bidelta network. While there is an exponential number of nonequivalent Delta networks, all $N = n^k$ node, k-stage bidelta networks configured with $n \times n$ switches are isomorphic to each other, and thus both topologically and functionally equivalent [331].

3.2.1.1   Shuffle-Exchange and Omega Network

If a MIN uses the same topology in all stages, it can be built as a single-stage, recirculating network with memory available at each switch. The 1-stage recirculating shuffle-exchange MIN shown in Figure 3.4 can pass data from any of N input ports to any of N output port in at most $\log_2 N$ steps. Although cheaper than an Omega network (introduced in this Section), the recirculating network cannot be circuit-switched, unless it is augmented with proper buffering. Moreover, this network has a large N/n bisection width, and can realize all permutation in $2 \log N - 1$ passes. One can reduce the number of passes by increasing the number of network stages.

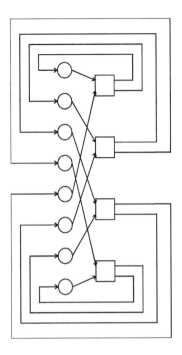

FIGURE 3.4: An 8-node single-stage shuffle-exchange (circles show PEs)

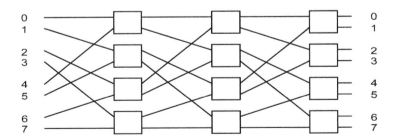

FIGURE 3.5: The Omega network for $N = 8$

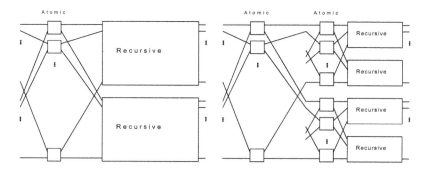

FIGURE 3.6:   Recursive construction of the Omega network

If we increase the number of stages to $\log_2 N$, we obtain the Omega network which has a unique input-output path for all input-output combinations [204]. An example of this popular MIN (all switches are of size 2) is shown in Figure 3.5.

Omega networks are representatives of highly symmetric, blocking MINs and can be easily partitioned to independent parts to support multitasking. Another convenient property is the source-based routing capability. Notice that Omega networks can also be constructed recursively, based on 2×2 switches. An N input Omega MIN is obtained from two $N/2$ input Omega networks, together with an extra switching stage added in front which distributes the N input ports, as shown in Figure 3.6.

3.2.1.2   Butterfly Network

A butterfly network with 16 nodes is shown in Figure 3.7. The butterfly network is based on a permutation which defines a different connection pattern for each stage of the network; this pattern essentially corresponds to the computation pattern of a one-dimensional FFT [208].

Butterfly networks are essentially Omega networks drawn differently, i.e. networks are wide-sense topologically equivalent. Actually, all bidelta MINs, including the Omega, butterfly, baseline, reverse-exchange, indirect binary cube, and their inverted versions, are topologically and functional equivalent.

An important technique improving performance over unique or limited-path blocking networks replaces buddy groups of switches with larger switches which offer better performance. This approach leads to high radix networks. Another approach creates multiple paths, either by providing extra switches or by chaining switches at each stage.

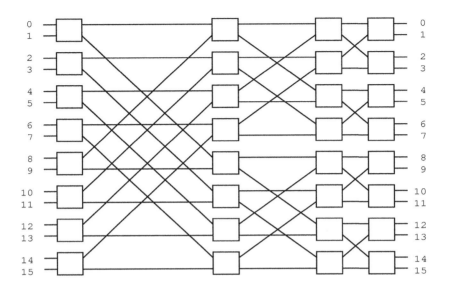

FIGURE 3.7: The butterfly network for $N = 16$

3.2.2 Permutation and Nonblocking Networks

3.2.2.1 Beneš Permutation Network

Rearrangeable networks (also called permutation networks) realize edge-disjoint paths to admit any arbitrary permutation. Permutation requests arrive in batch, and each batch is realized independently of previous connections, by establishing a new set of edge-disjoint paths. This may involve tearing down and rebuilding any of the existing connection paths.

The Beneš permutation network takes its name from Beneš who discovered it in 1962 [33] [30]. It is a cost-optimal permutation network, i.e. it has an asymptotically optimal number of switches, although setting these switches is essentially a sequential process that follows the recursive construction [377]. In Figure 3.8, we show a 16-input, Beneš network, designed by placing in tandem a butterfly together with its inverse, collapsing the two identical intermediate levels. Beneš network can also be recursively constructed from a more general three-stage Clos network (see next Section).

The Beneš network is topologically equivalent to several other networks, since the butterfly (or inverse butterfly) can be replaced with an (inverse) Omega, baseline, reverse-shuffle, indirect binary cube, or other bidelta MINs.

Another rearrangeable network is the $N = 2^n$-node, $2n - 1$-stage shuffle-exchange network. This multistage network is formed by concatenating two

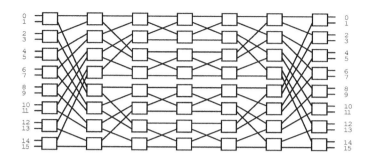

FIGURE 3.8:   Beneš permutation network for $N = 16$

Omega networks and collapsing the two identical intermediate levels [1]; a simpler routing algorithm is provided in [308].

3.2.2.2   The Clos Family of Networks

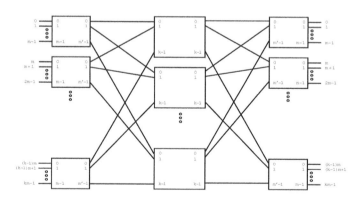

FIGURE 3.9:   A general, three-stage Clos network $(k = N/m)$

Clos networks were first defined by Clos in the 1950s [75]. The family of Clos networks of size N, which as a special case contains the Beneš permutation network, is constructed using the following rules (see Figure 3.9).

- N/m switches of size $m \times m'$ are used at input stage I.

- m' switches of size $\frac{N}{m} \times \frac{N}{m}$ are used at intermediate stage II.

- N/m switches of size $m' \times m$ are used at output stage III.

- The jth output link of every switch in stages I and II is connected to the jth switch at the following stage.

Optimized Clos networks can be constructed by expanding recursively an intermediate stage, and provide a framework for designing blocking, rearrangeable, and nonblocking MINs, by directly relating combinatorial power to the relative values of m and m'.

- Any three-stage Clos network with $m' < m$ is blocking.

- Any three-stage Clos network with $m \geq n$ is rearrangeable. The Beneš permutation network can be obtained from Clos networks by selecting $N = m \cdot r$, b) $m' = m$, and c) for the binary case, $m = 2$.

- Any three-stage Clos network with $m' \geq 2m - 1$ is strict-sense nonblocking; an example of a strict-sense nonblocking Clos network uses $N = 36$, $n = 6$, and $m = 11$.

Other nonblocking networks, including Cantor [58], multi-butterfly [22] [19], multi-Beneš [205] [315] [316], and randomly wired multistage networks [219] [223], are based on stacking (or cascading) simple multistage networks, thus, they are expensive for on-chip implementation. Furthermore, generalized connectors, used in multicasting or multiaccess memory networks, realize all input-output relations (not just permutations) through edge-disjoint paths, but require complex routers, e.g. broadcasting switches [391] [256] [355] [72].

Thus, nonblocking Clos-based MINs have good performance and nice topological properties, such as symmetry, robustness, small degree, small diameter, and large bisection width. Thus, despite a rather large network cost and extendibility, small locality, increased wire density (with many long interconnection wires), and rather expensive VLSI layout, they can be considered for NoC realization in data parallel applications, such as multimedia processing.

3.3 Mesh and Torus

Despite a relatively large diameter and average distance, regular, low-dimensional, point-to-point packet-switched topologies are more cost-effective for NoC realization. In particular, they offer more uniform wire density, especially for non-uniform, hot-spot traffic. Furthermore, in contrast to multistage networks, these topologies exploit locality by connecting each router to one or more NoC resource network interface(s).

Meshes and their direct generalizations, e.g. n-dimensional meshes or tori, are common examples of point-to-point topology-dependent NoC solutions due to their regularity, scalability, and conceptual simplicity. Several 2-dimensional mesh variants exist. The predominant network topology in the parallel interconnection literature is a 2-dimensional $n \times n$ mesh. This topology consists of $N = n^2$ nodes (usually PEs), arranged in a two-dimensional $n \times n$ grid. Each node is connected to its (at most) four neighbors. Nodes can be identified by their coordinates in the grid: the node at position (i, j), $0 \leq i, j < n$, is denoted by $P_{i,j}$, where position $(0, 0)$ lies in the upper left corner (as in a matrix). All definitions carry on to nonsquare $m \times n$ meshes. The 2-d mesh has nice topological properties, including simple VLSI layout, especially for a homogeneous Multicore implementation with identical cores. Usually the adopted routing scheme is the simple deterministic XY routing, resulting in simple and efficient router design. Several years of research have addressed efficient routing methods based on alternate paths, randomization, adaptivity, and fault tolerance. A plethora of interesting mesh embedding properties has been proposed for realizing efficiently, and in a scalable manner, several important platform-independent communication or computation patterns arising in parallel applications, such as image processing, linear algebraic computation and numerical solution of partial differential equations. Notice, however, that unlike data parallel applications, 2-d mesh may not be the most appropriate topology for irregular instruction-level parallelism, e.g. in digital signal processing or data flow processing applications.

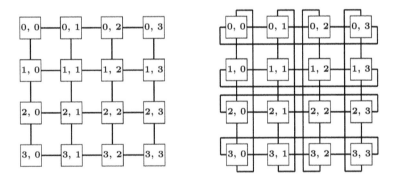

FIGURE 3.10:   A 4×4 mesh on the left, and a 4×4 torus on the right

The *2-dimensional torus* further reduces the 2-d mesh diameter by adding wrap-around links. These links run between $P_{i,n-1}$ and $P_{i,0}$ for all $0 \leq i < n$, and between $P_{n-1,j}$ and $P_{0,j}$, for all $0 \leq j < n$. These definitions are illustrated in Figure 3.10. A major benefit for tori compared to mesh is smaller diameter ($2\lfloor n/2 \rfloor$ vs. $2n - 2$) and larger bisection width ($2n$ vs. n); hence, as a consequence, the maximal speedup for algorithms without much "locality" is limited to n, with n^2 nodes. Moreover, notice that tori have higher structural uniformity, since they are vertex- and edge-symmetric, i.e. there are no corners that have to be treated specially. Since the $n \times n$ torus can be embedded onto the $n \times n$ mesh with a dilation 2 embedding, i.e. nodes adjacent in the torus are at distance two in the mesh, any all port torus communication algorithm can be simulated on a mesh with delay factor 2.

3.4 Chordal Rings

Most constant degree topologies discussed in this Section can be derived from the well known family of Cayley graphs. This rich family of graphs can be used to generate small degree, low diameter networks. Cayley graphs are based on algebraic group theory, e.g. the N-node permutation group with composition operator, or integer with add/multiply operator. Nodes of the Cayley graph are all group elements, while edges (and therefore routing) are based on applying the group operator function (called generator, $\otimes$), i.e. node x connects to node y, if and only if $x \otimes \gamma_i = y$ for some $\gamma_i \in S$. Cayley graphs share many nice topological properties. For example, all Cayley graphs are vertex symmetric, and many are also edge-symmetric if all operator pairs are related through a group automorphism. Moreover, almost all Cayley graphs are Hamiltonian, and many are hierarchically recursive and optimally fault tolerant.

Circulant networks (also called circulants) are special Cayley graphs and digraphs defined on cyclic addition groups Z_N, where N is a positive integer greater than or equal to 3. Circulants arise in the study of periodic or multiply symmetric dynamical systems and have circulant adjacency matrices in which each row after the first is a cyclic shift of its previous row. In fact, the study of circulant networks dates back to the 1940s in the investigation of periodic or multiple symmetrical systems with application in the theory of crystal structures (see [294]).

Next, circulants are formally defined. Let $G(N; s_1, s_2, \cdots, s_k)$ be the digraph with N nodes, labeled with elements from Z_N, and each node x is adjacent to k other nodes $x + s_i$, $i = 1, 2, \cdots, k$. The graph $G(N; \pm s_1, \pm s_2, \cdots, \pm s_k)$ is the undirected version of $G(N; s_1, s_2, \cdots, s_k)$ where each node x is adjacent to $2k$ other nodes $x \pm s_i$, $i = 1, 2, \cdots, k$; naturally, $s_1, s_2, \cdots, s_k$ are called skip distances. The addition or subtraction is taken modulo N.

In this perspective, chordal rings can be defined as circulant graphs with $s_1 = 1$, i.e. with a chord of length 1. In addition, circulants with $k = 2$ are called double loop networks, that are a special case of chordal rings; notice that the edge bisection of any chordal ring is upper bounded by $2(s + 2)$ [187].

For all circulant graphs of degree 2 and 3, there exists a simple planar

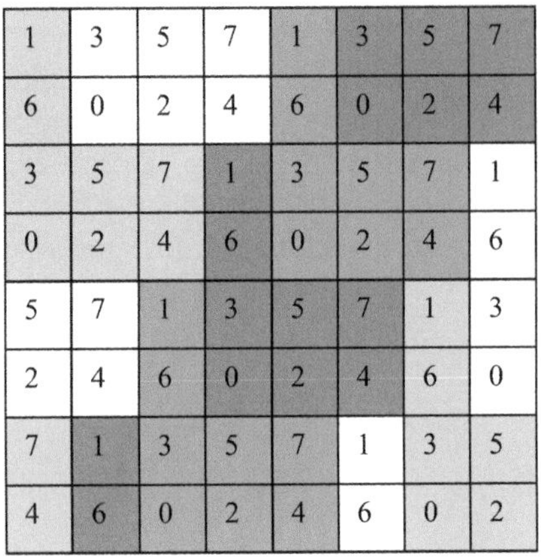

1	3	5	7	1	3	5	7
6	0	2	4	6	0	2	4
3	5	7	1	3	5	7	1
0	2	4	6	0	2	4	6
5	7	1	3	5	7	1	3
2	4	6	0	2	4	6	0
7	1	3	5	7	1	3	5
4	6	0	2	4	6	0	2

FIGURE 3.11:   Planar representation of undirected circulant $G(8; 2, 3)$

geometrical representation. For this representation, the Euclidean plane sub-divided into square tiles. Each tile represents a vertex of the graph, with the tile at $(0, 0)$ representing the vertex 0. For any tile representing vertex u, the tile to the right represents $u + s_1$, while the tile above it represents the vertex $u + s_2$. Thus, the s_1 edges correspond to connections between pairs of horizontally adjacent tiles, while the s_2 edges correspond to vertically adjacent tiles. The representation of the vertices in the plane repeats periodically and all graph vertices are contained in a rectangular or L-shape basic tile [215] [387]. The shape of this basic tile containing all nodes defines a minimum distance diagram [67]. Figure 3.11 illustrates the basic tile for planar representation of the undirected circulant $G(8; 2, 3)$. For vertex- and edge-symmetric graphs, this diagram can be used for obtaining the diameter, and for designing efficient broadcasting algorithms, grid embeddings, and planar or multi-planar VLSI design layouts. The L-shape concept also extends to hyper L-shapes for circulant graphs of higher degree [157].

We next consider optimal chordal ring topology selection. Notice that several tradeoffs must be considered, since a single NoC topology cannot optimize at the same time many performance metrics, e.g. diameter, average distance, or bisection width. Despite this problem, through theoretical graph exploration, we can identify interesting chordal ring families which provide nice topological properties. Next, we examine dense degree 2, degree 3 and degree 4 chordal rings which attempt to optimize their diameter for a given number of nodes.

Optimized directed 2-circulant chordal rings, or double loop networks can achieve near optimal $O(\sqrt{N})$ diameter provided that a complex routing algorithm is implemented, requiring extra pre-processing and possibly use of routing tables; notice that in general, the optimal diameter of a chordal ring with degree d and N nodes is $dN^{1/d}$ [264]. Moreover, chordal rings of degree 2 have nice topological metrics if the skip distance s is selected properly.

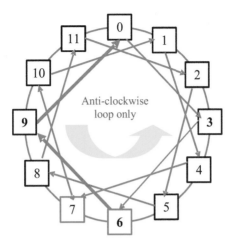

FIGURE 3.12: 12-node directed 2-circulant with $s_1 = -1$, $s_2 = \lfloor \sqrt{N} \rfloor = 3$

Among degree-2 graphs, the 2-circulant directed graph with $s_1 = -1$ (representing backward links), and $s_2 = \lfloor \sqrt{N} \rfloor$ (representing forward links), appears particularly interesting, since it combines a small diameter (approximately $\sim 2\lfloor \sqrt{N} \rfloor - 2$) and average distance (approximately $\sim \lfloor \sqrt{N} \rfloor - 1$) with a simple shortest-path routing algorithm and near optimal fault tolerance [34] [156] [157]. Figure 3.12 illustrates an instance of this family with $N = 12$, $s_1 = -1$ and $s_2 = 3$. In this case, the diameter is 5 ($< 2\lceil \sqrt{N} \rceil - 2 = 6$) and the edge bisection width is 8. It should be noted that the obvious choice $s_2 = \lfloor \sqrt{N} \rfloor$ is not always optimal, e.g. the diameter of $G(16; -1, \lfloor \sqrt{N} \rfloor = 4)$ is 6, while that of $G(16; -1, 7)$ is 5.

Nevertheless, the proposed 2-circulant has a simple deterministic shortest-path routing algorithm which combines forward and backward steps in order to reach its destination. The algorithm first routes packets forward not overshooting the destination, and then uses backward links to reach the target node. Since forward and backward steps can be taken in any order, this routing algorithm has high adaptivity.

Simple and optimal broadcast, multicast and permutation routing can be based on different L-shape representations, i.e. the planar geometrical representation of 2-circulants discussed previously [67] [102] [155] [387]. For example, the permutation routing algorithm a) first routes packets using a number of long steps, and then b) routes packets using short steps, assigning higher priority to the packet that takes the longest distance (selected using a required

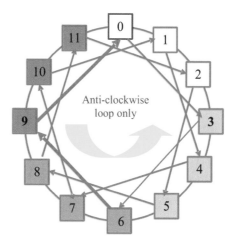

FIGURE 3.13:   Multinode accumulate algorithm on 2-circulant

packet counter). Similarly, other intensive communication algorithms can be designed.

For instance, an optimal routing algorithm for multinode accumulate (corresponding to many-to-few routing and partial or all-reduce operations) is as follows; multimode accumulate is the dual operation to multimode broadcast. Since the graph is vertex symmetric (see Figure 3.13), without loss of generality, we can consider receiving node 0. Therefore, the algorithm routes packets from half the nodes (those numbered from 0 to $N/2 - 1$) through their backward links. Packets from remaining nodes at first propagate using their backward links to their nearest primary node, numbered as a multiple of $\lfloor \sqrt{N} \rfloor$, and then move towards node 0 using forward chords. Notice that node 0 receives approximately the same number of packets from each direction, thus resulting in small queues and small delay. This algorithm can be easily extended to cover many-to-few patterns.

Notice that another interesting degree 2 graph, similar to the above circulant graph, is the 2-d $m \times n$ torus, with unidirectional links in both dimensions and diameter ($\lceil m/2 \rceil + \lceil n/2 \rceil$).

Chordal rings of degree 3 or 4 often resemble a mesh or torus. For example the ILLIAC IV parallel interconnect, often described as 8×8 mesh or torus, is an undirected 64-node chordal ring with skip distances ($s_1 = \pm 1$, $s_2 = \pm 3$). Figure 3.14 shows a reduced version of ILLIAC IV for 9 nodes.

In general, chordal rings of degree 3 are less symmetric than chordal rings of even degree, since they are Cayley graphs on a non-commutative group. Complex theoretical research has concentrated on deriving diameter and sometimes bisection bounds. Notice that although chordal rings reduce diameter to even less than that of a mesh or even a torus, they are not optimal in terms of

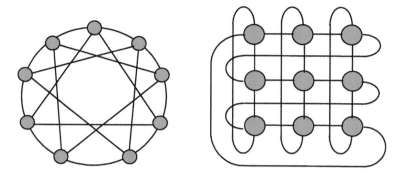

FIGURE 3.14: Equivalent representation of degree 4 ILLIAC IV topology

diameter, with respect to the upper bound of $3N^{1/3}$.

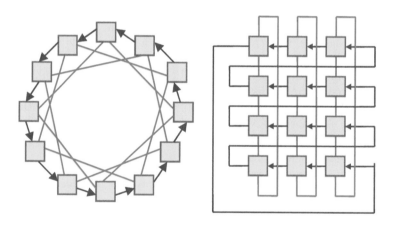

FIGURE 3.15: Equivalent representations of 12-node 3-circulant with $s_1 = -1$, and $s_2 = \pm\lfloor\sqrt{N}\rfloor$

A general family of 2-d mesh-like chordal rings can be defined with $N = kxl$ nodes, where $l = \lfloor\sqrt{N}\rfloor$. Figure 3.15 illustrates a 12-node degree-3 directed chordal ring obtained with $k = 4$, and $l = \lfloor\sqrt{N}\rfloor = 3$. This mesh-like representation can provide certain competitive topologies for $N \leq 120$. Notice that this family admits a simple routing algorithm with two phases. At first, we can locate the k-loop containing the destination node in at most k steps using the unidirectional loop (blue edges). Then, we can locate the target node using the bi-directional loops (red edges). In this case, the diameter is: $(l - 1) + \lfloor k/2 \rfloor$.

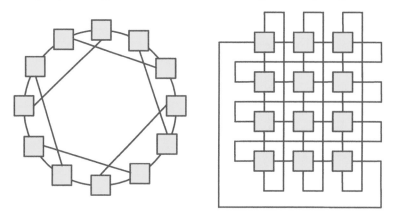

FIGURE 3.16:   A 3-circulant with $s_1 = \pm 1$ and $s_2 = \lfloor \sqrt{N} \rfloor$ for odd N or $s_2 = -\lfloor \sqrt{N} \rfloor$ for even N

Two other interesting N-node 3-circulants have a) $s_1 = -1$ and $s_2 = \pm \lfloor \sqrt{N} \rfloor$, and b) $s_1 = \pm 1$ and $s_2 = \lfloor \sqrt{N} \rfloor$. These graphs can be defined for any N, and achieve almost $1.5 \lfloor \sqrt{N} \rfloor$ diameter, with a fairly simple routing algorithm, similar to Spidergon (see Chapter 4). Similar diameter ($\sim 1.5\sqrt{N}$) is obtained by the unidirectional x-direction and bidirectional y-direction 2-d torus, while even smaller ($3N^{1/3}$) is obtained by the 3-d unidirectional torus. Finally, notice that the 3-circulant shown in Figure 3.16, which is defined with $s_1 = \pm 1$, and $s_2 = \lfloor \sqrt{N} \rfloor$ for odd N, or $s_2 = -\lfloor \sqrt{N} \rfloor$ for even N, despite its apparent complexity has approximately $\sim 1.5\sqrt{N}$ diameter and an efficient tree-based routing algorithm.

Finally, an interesting topology is the degree 4 undirected chordal ring with $s_1 = 1$, and $s_2 = 2k + 1$, and $N = 2k^2 + 2k + 1$, $k > 1$ nodes. In this case, the diameter is k and a fairly simple routing algorithm can be used (similar to the one provided for the degree-2 circulant graph). This corresponds to a 70% reduction in diameter (and slightly more in the average distance) over a 2-d torus with similar size, degree and number of links. This family of chordal rings is also dense, i.e. optimal in the sense that it has the minimal diameter among all possible circulants of degree four. Moreover, all optimal 4-circulant graphs with diameter k are isomorphic to this chordal ring. Optimal diameter and adequate routing and deadlock-avoidance mechanisms translate into better system performance when typical parallel applications are executed over a multiprocessor SoC. Finally, notice that although this graph has limited extendibility, it has many nice properties, such as large edge bisection, vertex- and edge-symmetry, a Hamiltonian cycle, high fault tolerance, and efficient point-to-point routing and broadcasting based on L-shape embedding.

Finally, **Periodically regular chordal rings (PRCR)** separate the ring into several groups and provide skip links for members of each group that

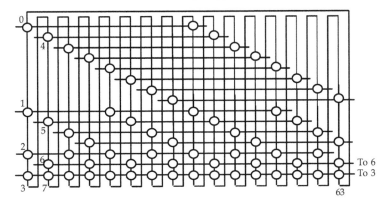

FIGURE 3.17: VLSI layout for a 64-node PRCR [264]

lead to the same relative position in given symmetrically-defined destination groups [264]. In addition, PRCRs may remove certain chordal links for better cost-performance, i.e. they trade node-symmetry for a small diameter. For degree k, the PRCR diameter is at most as a k-dimensional mesh. These graphs offer simple modified greedy routing by routing first route to the head of a group, and then using pure greedy routing. They also have nice VLSI layout properties, as they can be embedded on the plane with a small number of crossovers. An example graph is shown in Figure 3.17.

3.5   Other Constant Degree Topologies

Next, a very brief survey of the most remarkable theoretical results in other low-cost (constant degree), high performance (small diameter) graphs, such cube-connected cycles, star-connected cycles, fat trees and de Bruijn network will reveal interesting families of constant degree graphs as viable alternatives to chordal rings for low cost, high performance prospective NoC topologies. Moreover, most of these graphs feature a large bisection width and nice communication properties.

The n-dimensional binary hypercube (also called n-cube) is a static interconnection network with $N = 2^n$ processors, numbered $0, 1, \ldots, N - 1$. A processor is addressed by a binary vector $x = x_1 x_2 \ldots x_n$, where $x_i \in \{0, 1\}$, for all $1 \leq i \leq n$. Two processors with addresses x, y are connected if their binary representations differ in exactly one bit. By examining the butterfly topology (e.g. Figure 3.7), we can see that each node of the n-cube corresponds to a collapsed row in the 2^n-node butterfly network [208]. Both the degree and diameter of the n-dimensional binary hypercube equal $\log N$, while its bisection width is $N/2$, and the graph is $(2n - 1)$-connected.

The hypercube topology can be recursively partitioned into smaller hypercubes, hence divide and conquer algorithms can be efficiently implemented on hypercubes. For example, the n-cube can be decomposed into two $(n - 1)$-cubes, or one $(n - 1)$-cube and two $(n - 2)$-cubes. Partitioning is based on grouping together hypercube nodes with identical bit patterns at given dimensions.

Important topological characteristics of the hypercube include its large number of available node- and edge-disjoint paths, efficient shortest path routing for common algorithmic patterns, and rich embedding properties [208].

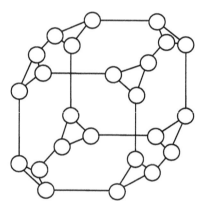

FIGURE 3.18:   A 4-dimensional cube-connected cycles

The degree 3, n-dimensional, $n\,2^n$-node **Cube-Connected Cycles (CCC)** is derived from the n-cube by replacing each hypercube node with an n-node ring. Thus, CCC is a subgraph of the $n + \log_2 n$-cube. The n-dimensional CCC of $N = n\,2^n$ nodes has $2n-1+\lfloor n/2 \rfloor$ diameter, $\sim 7n/4$ average distance and $N/2$ bisection width. The graph is not hierarchically recursive, but it is Hamiltonian, vertex- and edge-symmetric and it can emulate a hypercube step in just $\log_2 N$ steps. Actually, it can emulate all normal (ordered dimension) hypercube algorithms with constant dilation.

The n-dimensional **star network** is a Cayley graph with $N = n!$ nodes. Each node is labeled as $x_1 x_2 ... x_n$, a permutation of $\{1, 2, ..., n\}$. For each i, node $x_1 x_2 ... x_i ... x_n$ connects to $x_i x_2 ... x_1 ... x_n$, through the dimension i link (notice that x_1 and x_i are swapped). An example is provided in Figure 3.19. For an n-dimensional Star network, the degree is $n-1 = O(log N/log log N)$, its diameter is $\lfloor 3(n - 1)/2 \rfloor$ and its bisection width is $N/2$. In fact, the diameter of the Star is asymptotically optimal in respect to diameter [378]. It also has

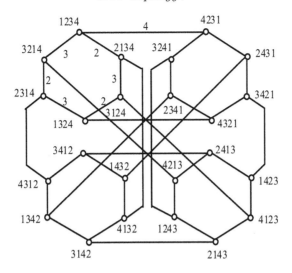

FIGURE 3.19: The 4-dimensional Star network

a simple non shortest-path routing algorithm that takes $2n - 3$ steps.

As shown in Figure 3.20, the $(n - 1)n!$ node **Star-Connected Cycles** (SCC) can be derived from n-dimensional Star network by replacing each node with an (n–1)-node ring. This graph has degree 3, diameter $2\lfloor 3(n - 1)/2 \rfloor - 1 + \lfloor n/2 \rfloor$ and a large, but unknown bisection width. Although the SCC diameter is similar to a corresponding size CCC network, routing and intensive communication algorithms for SCC are more complex.

Another proposed NoC topology is the **Fat Tree** [269] [210]. A k-ary fat tree is a modified tree structure with an increasing channel bandwidth as we move from the tree leaves (allocated to NoC resources) towards higher levels in the tree near the root (configured using router elements). Thus, a fat tree has a logarithmic diameter $(2log_k N)$ and a very large bisection, thus avoiding packet congestion which is typically present near the root due to packets crossing different tree sub-networks. Furthermore, routing is simple, i.e. routing upwards towards the most common ancestor, and then downwards towards the final destination

Although fat trees have exponential network extendibility, they also have a large bisection width and small diameter, competitive for small to medium size NoCs. For example, as shown in Figure 3.21, a 64 leaf, 2-level fat-tree organized with 4 16×16 switches at the top level and eight 8×8 switches at level two has 64 edges and bisection width 32. Examples of networks that employ a fat tree topology include the parallel CM-5 supercomputer [208] and the SPIN NoC [132].

The concept of a de Bruijn digraph was originally defined in 1946 in [92]. The n-dimensional k-ary de Bruijn network, denoted as $D(k, n)$, is modeled as

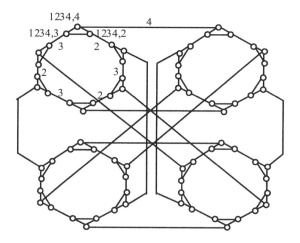

FIGURE 3.20:   The 4-dimensional SCC network

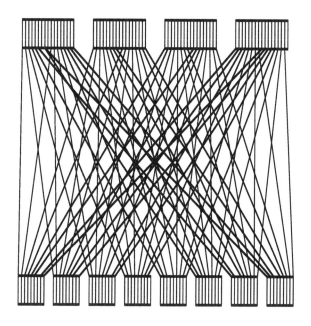

FIGURE 3.21:   A 2-level Fat tree with 64 nodes

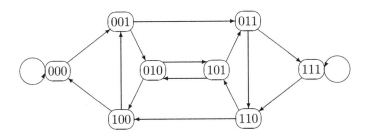

FIGURE 3.22:   de Bruijn graph $D(2,3)$

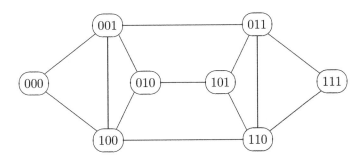

FIGURE 3.23:   de Bruijn graph $UD(2,3)$

a directed graph (called digraph) with k^n nodes. Each node is labeled as an n-bit k-ary number $a_1 a_2 \cdots a_n$ where $a_i \in [0, \ k-1]$. Node $a_1 a_2 \cdots a_n$ is adjacent to $a_2 a_3 \cdots a_{n-1} \alpha$, where $\alpha \in [0, \ k-1]$. Each node in $D(k, \ n)$ has indegree and outdegree equal to k. Self-loops do exist at certain nodes. For example, in $D(2, 3)$, node 000 (and node 111) is adjacent to itself. The undirected version of $D(k, \ n)$, where each link (a, b) is changed to an undirected edge $\{a, \ b\}$ and self-loops and multiple edges are deleted, is denoted as $UD(k, \ n)$. The graphs $D(2, 3)$ and $UD(2, 3)$ are shown in Figures 3.22 and 3.23, respectively. Notice that $B(d, \ k)$ is not vertex-transitive. $B(d, \ k)$ has dk nodes, indegree and outdegree d, and diameter k, and bisection width $2^n / n$. Simple non-minimal routing from $x_1 \ldots x_k$ to $y_1 \ldots y_k$ can be obtained by following the route: $x_1 \ldots x_k \rightarrow x_2 \ldots x_k y_1 \rightarrow x_3 \ldots x_k y_1 y_2 \rightarrow x_k y_1 \ldots yk - 1 \rightarrow y_1 \ldots y_k$.

Alike butterfly and CCC, de Bruijn networks are constant degree graphs that can emulate the N-node hypercube on many computational tasks with only N nodes (instead of $N \log_2 N$ nodes for the CCC and butterfly). The N-node binary de Bruijn network has $\log_2 N$ diameter and admits cycles of all different lengths, i.e. pancyclic [34] [147] [153]. Moreover, it contains an $(N-1)$-node complete binary tree.

Theoretical approaches for low-cost, high performance network design attempt to reduce the degree of an arbitrary large network of constant diameter at a rate faster than the corresponding increase in diameter or reduce the diameter of an arbitrary large network of constant degree at a rate faster than the corresponding increase in degree. More practical techniques focus on considering the effect of channel width. The exponential increasing bandwidth per link in on-chip routers can be more effectively utilized by increasing the radix or degree of router nodes than by making channels wider. Thus, as shown in Figure 3.24, **high-radix routers** essentially combine together independent stages of a buffered multistage network. These hierarchical structures do not affect network bisection bandwidth, but act as concentrators which aggregate traffic, thus improving resource sharing, performance and scalability.

High-radix micro-architectures signify a shift from conventional low-radix topologies, such as 2- and 3-d torus or mesh. In fact, a high-radix (64) router has been implemented in the Cray YARC router used in the modified folded-Clos interconnection network of the Cray BlackWidow supercomputer which introduces extra links to connect directly neighboring subtrees [304]. This network offers twice the performance of a comparable-cost Clos network on balanced traffic.

NoC topologies based on high-radix routers include the concentrated mesh and flattened butterfly. The concentrated mesh connects several cores at each router [24], while the flattened butterfly topology combines routers in each row of the conventional butterfly topology, while preserving inter-router connections [184]. These topologies reduce cost (due to a smaller number of internal channels and buffers) and network diameter, and offer higher bandwidth, lower latency, and energy-efficiency.

Assuming similar network bisection bandwidth, the flattened butterfly has

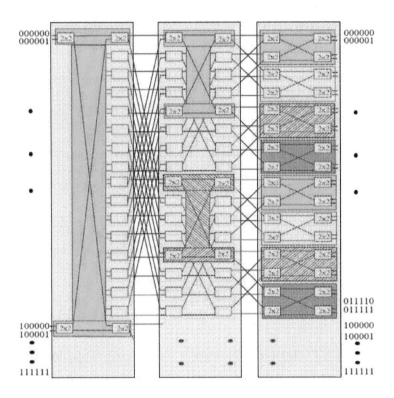

FIGURE 3.24:   A concentrated (high radix) network based on a butterfly

higher wiring complexity which can be reduced by inserting repeaters and pipeline registers, but improves node bandwidth by 50%, latency by 28%, and power consumption by 38%, compared to simple 64-node 2-d mesh architecture. Flattened butterfly also offers bypass channels for connecting directly to next-level routers and efficient 2-d planar VLSI layout.

For balanced traffic, flattened butterfly approaches the cost-performance ratio of a butterfly and has twice the ratio of a Clos network. For communication patterns with transient load imbalance, non-minimal globally-adaptive routing on flattened butterfly provides the same ratio as a folded-Clos topology and an order of magnitude better performance than a conventional butterfly.

3.6   The Spidergon STNoC Topology

ST Microelectronics has proposed an innovative point-to-point interconnection topology, called Spidergon, which targets cost-performance tradeoffs compared to other NoC topologies. Spidergon has been defined in the framework of the NoC research activity at AST Grenoble Lab (Advanced System Technology) of ST Microelectronics.

The proposed Spidergon STNoC topology generalizes the previous circuit-switched ST Octagon network processor topology to a simple bidirectional ring, with extra cross links from each node to its diagonally opposite neighbor. The Spidergon topology overcomes major drawbacks in the Octagon and implements relevant NoC requirements.

This Section describes Spidergon, examining its topological properties and exhibiting its theoretical foundation within the family of chordal rings.

3.6.1   The ST Octagon Network Processor Topology

Octagon, proposed by ST Microelectronics [178], is an interesting circuit-switched interconnect based on a regular point-to-point topology designed for targeting the network processor domain. The basic configuration shown in Figure 3.25 is an eight node bidirectional ring with cross connections. Thus, each processor is directly connected to two adjacent nodes and the node directly across. This basic topology has small degree (3), diameter of just two hops, and allows for a simple and efficient shortest path routing. The network provides a high concurrency, low latency on-chip communication architecture, able to meet network processing needs. It has significantly higher performance than bus-based on-chip communication, while having less wiring complexity.

The ST Microelectronics Octagon has two main drawbacks that limit flexibility, efficiency and scalability as a prospective NoC architecture: circuit switching based on centralized arbitration, and significant network extendibil-

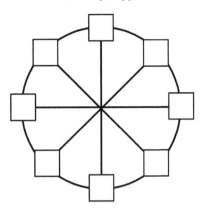

FIGURE 3.25:   The degree-3 Octagon topology

ity (8 nodes) which represents high granularity when scaling to a larger network configuration.

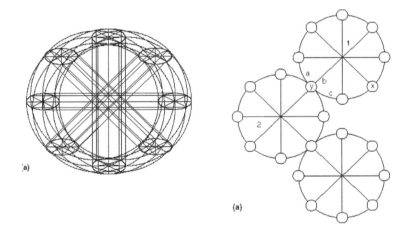

FIGURE 3.26:   The two Octagon network extendibility strategies

Figure 3.26 shows two strategies proposed for building larger, complex Octagon networks. Strategy A involves connecting 64 nodes as eight Octagon clusters visualized as a Cartesian product of basic octagon topologies with a processing element at each node; this solution has very small diameter, but high link cost [178] [177]. An alternative cost-effective strategy B connects Octagon clusters through so-called bridge nodes, indicated as "y". Bridge nodes are more complex than other nodes, and can be a bottleneck when application mappings require much inter-cluster traffic.

3.6.2   The Spidergon STNoC Topology

Figure 2.8 shows examples of NoC topologies that have been previously proposed. Each NoC topology offers a different set of tradeoffs in terms of metrics, such as

- vertex symmetry, which affects routing or scheduling cost and performance,

- network degree, which affects the operating frequency and the complexity of the routers,

- network extendibility, which should be as low as possible to maximize flexibility in Multicore design, and average distance between nodes.

Although these regular topologies offer many different theoretical research directions, in the Multicore domain the overriding consideration is the final price/performance ratio of the resulting chip. For this reason, ST Microelectronics developed the new Spidergon topology, which promises to deliver a good tradeoff between theoretical performance and implementation cost.

Notice that topologies with increased connectivity, such as 2-d mesh, provide very good theoretical metrics, but in most cases, these features cannot be fully exploited due to the nature of communication traffic in Multicore SoC applications. On the other hand, simple topologies, such as rings, are cost-effective in terms of manufacturing cost, but deliver relatively poor performance, especially as the number of connected cores increases.

The Spidergon STNoC topology is a regular, point-to-point topology similar to a simple bidirectional ring, except that each node has, in addition to links to its clockwise and counter-clockwise neighboring nodes, a direct bidirectional link to its diagonally opposite neighbor (see Figure 3.27).

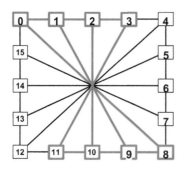 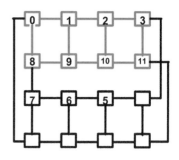

FIGURE 3.27:   Equivalent representation of Spidergon STNoC for $N = 16$

The Spidergon topology is vertex transitive and has constant network extendibility (equal to 2). The 8-node Octagon network is a member of the Spidergon topology. Spidergon, as constant degree vertex-symmetric network, offers simple, local and fast routing and scheduling decisions and simple router implementation.

Indeed from a routing point of view, any packet that arrives at a node which is not its final destination can be forwarded clockwise, anticlockwise or across the network to its diagonally opposite node. Moreover, the schematic Spidergon topology translates easily into a low-cost practical layout: the right side of Figure 3.27 shows a possible equivalent planar representation in which physical connections between nodes only need to cross at one point in the chip. However, in general, this simple placement is not the final on-chip physical layout of a 16-node STNoC, since the NoC would usually consist of heterogeneous IPs of different sizes and aspect ratios, disrupting regular, predictable positioning of the IPs within the network.

Formally, the Spidergon graph belongs to the general family of undirected circulant graphs. Despite the vast theoretical research on circulant graphs, double loop networks, and chordal rings, Spidergon has not been studied at all, since the main focus is usually on asymptotically efficient graphs targeting mainly massively parallel systems.

Within the class of circulant graphs, the Spidergon network connects an even number of nodes $N = 2n$, $n = 1, 2...$ as a vertex-symmetric 3-circulant with $k = 2$, $s_1 = 1$ and $s_2 = (l + n)mod N$. Thus, Spidergon consists of a bi-directional ring in both clockwise (right), and anti-clockwise (left) directions; in addition, for each node there is a cross connection, i.e. from node i, $0 \leq i < N$ to node $(i + n)mod N$.

The total number of edges in Spidergon is $3N/2$, and the graph diameter is $N/4$, its network extendibility is only 2 nodes, while its bisection width is limited (8 edges if $N = 4n$). Moreover, Spidergon topology allows for low-complexity identical hardware router implementation without routing tables and trivial, yet efficient, deterministic, shortest-path routing algorithms.

Compared to complex topologies, Spidergon offers a small number of links and simple implementation. For current, realistic NoC configurations with up to 60 nodes, the proposed Spidergon graph has a smaller number of edges and a competitive network diameter with respect to fat-tree or 2-d mesh topologies.

For example, the network diameter of a 4×5 mesh with 31 bi-directional edges is 7, while that of a 20-node Spidergon topology with 30 bi-directional edges is only 5. In general, Spidergon has a smaller diameter than 2-d mesh for network sizes up to 50. After 50 nodes the higher number of links of the 2-d mesh allows for better connectivity, although the difference is always smaller than 2 hops. As stated before, higher degree topologies, such as 2-d mesh or 2-d torus, do not deliver significant practical benefits in the Multicore domain for a variety of reasons, including reduced performance when the interconnect size leads to non-square (irregular) networks. Higher dimensional topologies improve performance metrics and topological properties for network sizes greater than 64 nodes. Thus, simple interconnect architectures are desirable, since they enable better NoC implementation, including higher operating frequencies and lower costs.

As discussed in Chapter 4, Spidergon STNoC interconnect technology supports a family of topologies based on the Spidergon, with a capability to customize the same router architecture depending on application requirements, e.g. by not instantiating certain links. Moreover, Spidergon STNoC supports aggregation and hierarchy which provide improved performance (e.g. reduced diameter) for a wide range of network sizes.

3.7   Comparisons based on Topology Metrics

NoC topology selection represents an important factor in overall NoC architecture performance and implementation. We have previously provided a thorough description of common network topologies along with a definition of theoretical performance metrics used to identify and compare graph properties.

Figure 3.30 presents a summary of static topological metrics for candidate NoC topologies: ring, Spidergon, $m \times n$ mesh, $k \times k$ torus and optimized degree 2 chordal ring. For all networks, we assume that two unidirectional links exist between any two nodes.

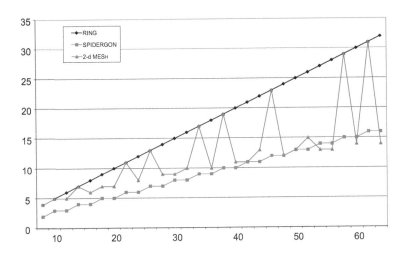

FIGURE 3.28:   Ring, Spidergon, and 2-d mesh diameter ($N = 8$ to 64)

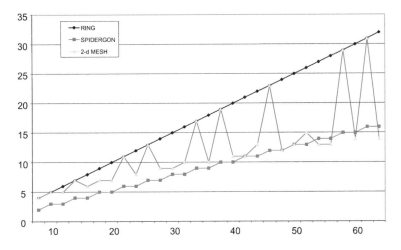

FIGURE 3.29: Ring, Spidergon, and 2-d mesh average distance ($N = 8$ to 64)

Based on Figure 3.30, Figures 3.28 and 3.29 compute diameter and average distance versus network size for three different topologies: ring, Spidergon, and 2-d mesh. These figures show that for $N < 60$, the Spidergon topology despite its constant bisection is very competitive. Due to its higher connectivity, the Spidergon topology always clearly outperforms ring in terms of diameter and average distance. Spidergon also competes favorably or outperforms 2-d mesh, although outcomes depend on network size. In fact, while Spidergon properties scale linearly with the network size, 2-d mesh behavior is quite irregular. This irregularity is a severe bottleneck in Multicore SoC design, since it complicates design space exploration in terms of cost-performance tradeoffs. For example, a 22 node 2-d mesh has smaller diameters and average distance metrics than a 24 node 2-d mesh. Although this issue may be solved in parallel systems by adding extra routers, i.e. implementing regular or square meshes, e.g. a 24 node mesh, it is a severe problem in application-specific SoC design where small design cost is fundamental.

From the above graphs, it also follows that 2-d mesh starts being more competitive in terms of diameter than Spidergon topology only for networks with a relatively large number of nodes, hence it is an unreasonable choice for near future Multicore platforms, where NoC structures will be implemented with only a few dozens of nodes.

In addition, as already stated, considering aggregation and hierarchical features, Spidergon topology can cope with system with a larger number of nodes. Therefore, Spidergon topology is an effective tradeoff solution for on-chip communication in next generation Multicores.

Property	Linear	Ring	Spidergon	2-d $m \times n$ Mesh	2-d $k \times k$ Torus	Chordal Ring
Symmetry	no	both	vertex	no	both	vertex
Degree	2	2	3	4	4	2
Links	$2N - 2$	$2N$	$3N$	$4mn - 2m - 2n$	$4N$	$2N$
Extendibility	1	1	2	$\min(m,n) \geq 2$ (constant rearrange)	$2k+1$	2
Node connectivity	1	2	3	2	4	2
Diameter	$N - 1$	$\lfloor N/2 \rfloor$	$\lceil N/4 \rceil$	$m + n - 2$	$2\lfloor k/2 \rfloor$	$\leq 2\lceil \sqrt{N} \rceil$
Average distance	$(3N-2)/4$ (if N even) $(3N^2 - 2N - 1)/(4N)$ (if N odd)	$N/4$ (if N even) $(N^2-1)/4N$ (if N odd)	$(2n^2 + 2n - 1)/N$ (if $N = 4n$) $(2n^2 + 4n + 1)/N$ (if $N = 4n+2$)	$(m + n)(mn - 1)/(3mn)$	$kN/(2N-2)$ (If k even) $k/2$ (if k odd)	$\sim \lceil \sqrt{N} \rceil$
Bisection width	1	4	8 ($N = 4n$) 10 ($N = 4n+2$)	$2\min(m,n)$ (if $\max(m,n)$ even) $2\min(m,n) + 2$ (if $\max(m,n)$ odd)	$4k$ (if k even) $4(k+1)$ (if k odd)	$\sim \lceil \sqrt{N} \rceil$ (for each direction)

FIGURE 3.30: Theoretical metrics for different topologies

Chapter 4

The Spidergon STNoC

4.1 Spidergon STNoC Interconnect Processing Unit

With multicore systems, the evolution from computation- to communication-centric platforms implies a novel perspective with respect to on-chip communication network infrastructures.

Thus far, on-chip communication infrastructures have relied on buses and crossbars that can be cascaded into several levels to eventually form large, multistage-like architectures. However, the complexity and time to fit these architectures to Multicore application requirements is not negligible. As stated in Chapter 1, novel on-chip communication infrastructures evolve towards flexible and programmable components defined as Interconnect Processing Unit (IPU). An IPU is essentially an on-chip communication network with hardware and software components which jointly implement key functions of different Multicore programming models through a set of **communication and synchronization primitives** and provide **low-level platform services** to enable advanced features in modern Multicore applications. Since synchronization primitives are usually implemented through atomic communication-based operations, next, we use the term communication primitives to denote in a broad sense both standard communication and synchronization primitives.

Although communication primitives and low-level platform services are concepts that play an important role in existing Multicore SoC, their binding within the context of on-chip communication networks into a new, unique entity called **IPU** is a novel idea. Currently there is no clear vision as to exactly where and how communication primitives and low-level platform services may be implemented. Low-level platform services are required for exposing available hardware through system software functions, including operating system calls, dynamic software libraries, and device drivers. These services could help address new market requirements and provide considerable savings in time-to-market for porting platform software.

As shown in Figure 4.1, the IPU is exposed to different programming models and middleware frameworks. Programming models are embedded in different parallel languages or programming environments. They are based on appropriate **language primitives** realized in terms of communication oper-

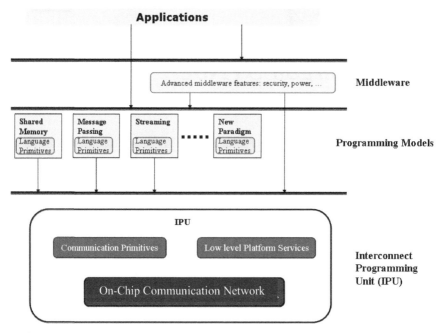

FIGURE 4.1:   A software view of an Interconnect Processing Unit

ations (e.g. read, write, send, or receive) and synchronization functions (e.g. test-and-set, or compare-and-swap) of the multicore architecture. Generally, language primitives are used by programmers to code applications.

Communication operations and synchronization functions can be implemented using two alternative approaches, either relatively simple, low-level communication primitives or very sophisticated ones. While simple primitives can support many different programming models through a possibly complex mapping phase, complex primitives provide efficient one-to-one mapping of a particular programming model at the risk of not being able to support other incompatible ones.

Communication operations and synchronization functions in traditional SoC or processor design have been strongly influenced by the shared memory programming model. This popular model relies on concurrent read/write with appropriately specified lock/unlock operations on memory locations for implementing typical shared memory programming primitives. These operations are mapped directly to load, store, and atomic communication primitives, e.g. swap, read-modify-write for STBUS, lock or exclusive lock for AXI, and read-exclusive, or read-linked for OCP. Simple and general-purpose shared memory communication primitives have been directly implemented in hardware, e.g. through two-phase request-reply on-chip bus protocols. For instance, a read operation will cause a load request on a physical shared memory location and a load reply will return the requested data. While a write operation produces a store request containing the data, a store reply contains an acknowledgement. Using shared memory communication primitives, it is possible to implement more complex languages primitives, such as send and receive. For example, synchronous blocking send and receive can be easily implemented using a 3-phase protocol based on a standard store communication primitive. More specifically, at first the sender can use a store primitive to inform the receiver, then, when ready, the receiver informs the sender by issuing another store primitive, and finally, the sender can directly send the data to the ready receiver by executing a final store primitive.

As discussed in Chapter 1, nowadays several start-up companies and universities propose new multicore programming models, such as streaming. Programming paradigms must cope with application complexity, providing efficiency and scalability of current and future multicore architectures. Thus, innovative programming model paradigms introduce new communication and synchronization mechanisms that are commonly implemented by runtime libraries or compilers built on top of common shared memory communication primitives. However, this approach leads to inefficient implementation of communication and synchronization mechanisms, degrading inter-processor communication performance, concurrency, and scalability. For this reason, we envision that IPU will exploit the lesson learned by the telecommunication community, providing the right communication primitives at the right abstraction level.

Low-level platform services are an important aspect of the IPU, since they provide the basic infrastructure for implementing efficiently application functionality of feature-rich devices. However, rewriting applications for different wireless platforms can present a daunting challenge for software vendors and device manufacturers. A practical way to overcome this problem is through an open software-based multimedia framework interfaces that provides hardware abstraction. Current platforms for mobile devices, such as Android [16] or LiMo [216], enable creation and integration of new application services through an operating system, and a modular plug-and-play middleware architecture (refer to Chapter 1).

For example, the Nomadik software framework [254] offers a high-level view (see Figure 4.2) which provides end-users with a set of uniform software interfaces that abstract the underlying processor design for a given operating system, and a low-level view which is a set of standard hardware interfaces for common application peripherals, such as LCD controllers, image sensors, or cameras.

In this context, middleware services that fit between the low-level API and the high-level client API can leverage on specific IPU low-level platform services. For example, an IPU can provide the basic plugs through which a high-level security framework can be built on top of a low-level platform service.

IPU software programmability, implies the capability of changing functionality at runtime by software and delivering new, enhanced end-user experi-

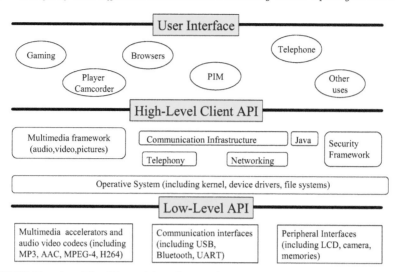

FIGURE 4.2:   The Nomadik software framework

ences. For instance, if a feature-rich camera device runs at a critical level of battery and the user still wants to take an important photo, e.g. during a birthday, sport, marriage, or theater event, thanks to IPU software programmability, it is possible to selectively power down idle components, keeping alive only the photo functionality, while also setting appropriate traffic requirements to safely store the photo into the memory. This feature can deliver a unique experience to the end-user, diversifying this device from similar products.

IPU software programmability is responsible for providing application flexibility and portability. However, if the IPU can also be configured at design time, then it can be optimized using specific Multicore SoC criteria, e.g. performance, power and chip area; refer to Chapter 6 for other interesting possibilities, such as static mapping of applications onto a specific NoC topology or runtime (dynamic) topology reconfiguration. Thus, alike configurable cores, e.g. from ARC [18], Toshiba (MeP) [360], Stretch [340], and Tensilica [354], IPUs are also characterized by hardware configurability. IPU hardware configurability involves pre-synthesis specification for on-chip communication infrastructure and opens new ways for optimizing communication. During this phase, it is possible to modify several architectural parameters, e.g. supported communication protocols and routing algorithms, packet and flit size, FIFO size, number of registers, and number of virtual channels.

Clearly, reconfigurable IPUs offer several advantages, such as co-design and simultaneous hardware and software optimization. For example, it is possible to avoid an IPU feature that does not capture system requirements, while at the same time, it is possible to add new IPU functions in a seamless fashion without a long development time or a prohibitive cost. Therefore, each IPU instance can be unique in terms of power, performance, area, and platform services. Semiconductor companies can use this intrinsic flexibility for providing a competitive product with a differentiation factor in the marketplace. However, verification of a configurable system is a major issue due to the myriad instances that a designer can generate.

For this reason, IPU design must be associated to an innovative EDA flow coupled with new methodology. This EDA flow should improve flexibility by introducing a certain degree of complexity which is hidden from the designer. Within this EDA flow, transaction- and cycle-accurate simulation models and corresponding profiling tools are very important. Transaction-accurate IPU models are especially useful in platform-level simulation, while cycle-accurate ones are important for obtaining qualitative and quantitative performance data for different configurability parameters. IPU profiling tools help analyze this data and obtain different statistical performance metrics.

Another important property is IPU extensibility which refers to the possibility of adding new low-level platform services and communication primitives that can simplify support of specific application requirements. These extensions can be either predefined or specified by the designer. Generally, IPU extensibility must rely on both software programmability and hardware configurability.

As a first attempt at designing a configurable IPU, Spidergon STNoC IPU technology consists of a flexible, pseudo-regular on-chip communication network implementing a set of customizable low-level platform services and a set of communication primitives.

The Spidergon STNoC IPU is a software programmable on-chip communication network that enables system designers to extend communication primitives. These communication primitive extensions are then automatically synthesized, placed, and routed into the Spidergon STNoC communication network. The communication network leverages graph properties of the Spidergon regular topology discussed in Chapter 3, while matching heterogeneity of Multicore SoCs through software programmability concepts (see also Section 2.3), application-specific hardware configurability and extensibility.

Existing NoC architectures are either based on a fixed regular network topology, or they are topology-independent, i.e. they can be customized to a particular application-specific communication graph. The Spidergon STNoC topology fills the gap between these two approaches, trading off regularity with customizability; for this reason, we call it a pseudo-regular topology. Using this concept, NoC topology becomes an architectural parameter that can be configured depending on the communication patterns exhibited by the application. Moreover, in order to address Multicore applications requirements for feature-rich devices, e.g. speech processing, video, GPS, security, and mobility, Spidergon STNoC technology provides a set of low-level platform services. The most important services defined in Spidergon STNoC are security, power management, and QoS. These services can be instantiated depending on the real target application and may be augmented by customer-specified services.

For example, by instantiating and using a security service, we can build a trusted Multicore system.

Since Multicore SoC application requirements become increasingly complex or even unknown until actual application use on the field, third party application developers for Multicore SoC platform push software and hardware to become more and more generic. This combination of hardware and software components into a standardized system brings about platformization of Multicore SoC, facilitating product differentiation only through high-level software functions.

Similar to configurable processor cores which provide an evolutionary improvement to standard processor cores, software programmable, configurable and extensible IPUs, such as Spidergon STNoC, are an evolution from traditional on-chip communication infrastructures. Considering the great success of configurable cores and the fact that hardware and software must become more generic, we envision that contemporary consumer-friendly and feature-rich devices will be based on configurable Multicore SoC platforms. These configurable platforms would combine IPUs (e.g. Spidergon STNoC) with a host processor (e.g. ARM) and several configurable cores (e.g. ARC, MeP, Stretch, and Tensilica) to easily handle multimedia contents (see Sections 1.1.1 and 1.2.1) and provide product differentiation through configurable hardware and software functions. Moreover, such platforms will increase productivity through hardware and software component reuse and provide product differentiation for a unique end-user experience.

4.1.1   The Spidergon Topology Family

As explained before, embedded multicore applications require customizable heterogeneous communication infrastructures. This heterogeneity can be captured through a network topology that allows configuring a hierarchy of subgraphs (or spanners) of the original topology.

Configurability of the Spidergon STNoC topology attempts to fill this gap between necessary regularity and heterogeneity through three main architectural and topological features.

- The Spidergon topology (refer to Chapter 3) can be customized and simplified depending on application traffic requirements.

- IP cores or subsystems can be connected to Spidergon STNoC in a flexible way, i.e. injection and extraction points can aggregate and combine traffic from potentially different IPs.

- Finally, regular networks can be replicated and connected to each other to compose hierarchies which reflect a functional composition of on-chip communication.

Figure 4.3 illustrates different families of topologies supported by the Spidergon STNoC. These topologies are essentially degree 2 or 3 Spidergon sub-

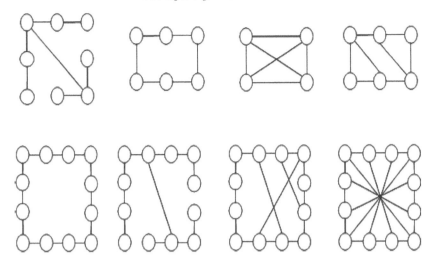

FIGURE 4.3: Different topologies supported by Spidergon STNoC

graphs that range from rings and simple spanning trees to irregular chordal rings. Depending on application traffic requirements and especially mapping of master and slave agents, connection paths can be removed if never used and cross connections customized to provide shortcuts between any pair of nodes in the ring. For instance, tree structures connecting many cores to one main destination can be easily embedded onto the Spidergon topology; a typical example is external DDR memory accessed by all connected IPs. This method results in a distributed tree equivalent to cascaded switches, but with the benefit of a regular network topology, i.e. several trees can be embedded simultaneously onto Spidergon, thus enabling multiple concurrent communication from processor or IP to memory.

The same theory, algorithms, and unique hardware logic support all Spidergon STNoC topologies. For example, routing schemes defined for the Spidergon topology apply to all derivative customizations.

Thus far, the topology has been examined as a graph, exhibiting routers and connections. Another fundamental issue in the Spidergon topology is the way subsystems of the Multicore SoC architecture are connected to this regular network. Spidergon STNoC enables the possibility to aggregate traffic at network injection and eventually split flows at ejection point. Aggregation is not only the way to capture and accommodate heterogeneity, but also to optimize on-chip network cost, bandwidth, and latency. Moreover sophisticated aggregation at network boundaries can associate physical independent flows to different virtual networks. Several heuristics are used to drive aggregration. The main principles relate to back-end locality (i.e. cores in the floorplan must be close to each other), logical locality (IP and processor cores must efficiently communicate each other), and application traffic characteristics, e.g. injection rate, complementary burstiness, and data bus size and frequency.

Spidergon STNoC architecture guarantees that each traffic flow is fully exposed to the communication services. Thus, from the architecture point of view, each flow corresponding to any connected IP or processor core can be tuned independently to the required service level, preserving IPU cost-efficiency and extensibility.

Selection of an on-chip communication architecture is extremely difficult due to several interrelated application software-related factors, such as increasing number of connected cores, interoperability of different socket protocols, and application traffic requirements. In particular, traffic may be characterized depending on the communication pattern of the application scenario into unknown, irregular or symmetric, random or deterministic, periodic, latency- or bandwidth-constrained, bursty, and short or long data transfer.

Since the on-chip network is physically distributed, the best approach to manage the above application complexity lies in functional separation of traffic. Due to this orthogonalization of flows, a global interconnect solution can be designed by either combining different physical networks in new application-specific hierarchical networks, or mapping different application flows onto the same physical infrastructure, eventually using virtual networks to guarantee application requirements.

Hierarchical network structures increase performance, since they reduce conflicts by exploiting locality, while ensuring global all-to-all connectivity.

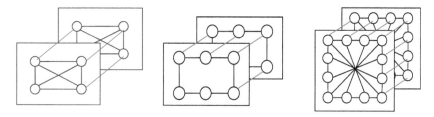

FIGURE 4.4:   Hierarchical topologies supported by Spidergon STNoC

Figure 4.4 shows a few examples of hierarchical networks supported by Spidergon STNoC technology. Further application-specific customization is possible, e.g. through hierarchical connections between network instances. Decoupling independent traffic flows for organizing nicely a SoC platform is not a new concept. In fact, many previously considered hierarchical on-chip communication architectures, such as custom segmented buses, or standard AMBA AHB or APB are based on separating different functional communications.

4.2   Switching Strategy

The switching strategy refers to how a packet traverses the route. More specifically, the switching technique determines how a message is fragmented and transmitted from an input of the on-chip network to an output resource by defining how channels and possible buffers along the path are allocated. There are four types of switching techniques:

- circuit switching,

- store-and-forward,

- virtual cut-through, and

- wormhole routing.

In circuit switching, point-to-point communication between source and destination involves two phases: circuit setup and message transmission. In the first phase, a physical path (circuit) from the source to the destination is reserved. Then, after a complete path is established through a series of switch allocation settings, the source can transmit the whole message directly to the destination at full bandwidth, i.e. end-to-end propagation delay is deterministic. Notice that messages may not be split into fixed size packets, but they can be directly transmitted as a continuous flow of bits. Duration of this path is associated to total message transmission time, or to a specific request to halt communication. The latter technique is advantageous for long messages, where the second phase is much longer than the first one.

In store-and-forward, each message is split into fixed-length packets. Each physical link has input and/or output buffers to hold at least one entire packet. Every packet is routed from source to the destination according to its address. A routing decision is made at each intermediate router only after the complete packet is stored at its input buffer. Store-and-forward is advantageous when messages are short and frequent, since each transmission results in at most one busy link in the packet's routing path. However, due to buffer implementation issues (memory versus flip-flop), and since the entire packet must be stored at each router, the relevant area may be quite large. Router chip area must be small for efficient NoC implementation; thus, using store-and-forward implies a strong limitation due to packet size. Network delay is another important drawback, since packet transmission cannot be pipelined. In fact, for packet routing, propagation delay is proportional to the product of packet size and distance, generally forcing designers to rely on complex adaptive routing techniques or high cost, low-diameter networks.

Virtual cut-through switching is a sophisticated and relatively expensive technique. Similar to store-and-forward, messages are split into packets with routers properly dimensioned to store entire packets received from each communication link. However, unlike store-and forward, packet information is pipelined into different segments (called flits) and routing decisions can be taken as soon as destination is available. Thus, if the physical link is idle and enough space is available at the next router, the first packet segment (header) containing the address information is immediately forwarded to the next router; subsequent packet segments are always directly forwarded on the route specified by the header, i.e. different packets cannot be interleaved over a physical link. Otherwise, similar to store-and-forward, incoming packet segments are stored in buffers at the current router until link and buffer resources at the next router or network interface become available. Notice that if no resource conflicts exist along the route, the packet is effectively pipelined through successive routers as a chain of segments. Thus, with low network congestion, transmission latency is greatly improved over store-and-forward and mainly depends on the sum of distance and packet size.

In wormhole routing packets are split into flits (flow control units) which are sent along the path opened by the header flit in a pipelined way [86]; at the physical layer, each flit may also consist of several smaller transmission units (usually a few bytes long), called physical units, or phits. With wormhole routing, router resources are normally dedicated to the packet until all flits have been transferred, and link transmission is based on flit flow control. While both store-and-forward switching strategy and virtual cut-through use per packet flow control, i.e. only the header carries flow control information, with wormhole routing each flit has its unique flow control [86]. This concept drastically reduces the amount of network buffering to a few flits, since routers do not have to store entire packets; the minimum buffer depth actually depends on link-level flow control and depth of packet pipelining. Moreover, similar to virtual cut-through, transmission latency with low network congestion is almost independent of the distance between source and destination [87]. However, since under heavy network contention blocked packet flits remain stored at the network interface and internal routers, wormhole routing is more susceptible to deadlock and congestion than virtual cut-through.

Spidergon STNoC adopts wormhole routing, which is nowadays commonly used in NoC design. Due to this switching technique, the Spidergon STNoC router has a simple architecture, occupies a small area, and is extremely fast in terms of operating frequency.

Transmission of different packets over a physical link cannot be interleaved or multiplexed freely without additional architecture support. In fact, if the header flit cannot proceed due to link contention, the whole worm of flits is stalled, occupying flit buffers in different routers on the currently constructed network path, while blocking other possible communications. Spidergon STNoC resolves this issue using virtual channels. This requires extra control logic and separate buffers that allow independent packet flows sharing the same physical link to be time multiplexed at flit granularity.

In general sense, flow control decides on the allocation of network buffer resources, as well as the related effects of packet stalling, dropping, or deflection. Thus, flow control refers to scheduling and conflict resolution policies during packet traversal, considering fairness and minimal congestion or reallocation. We distinguish between link-level (or hop-by-hop) flow control for managing flit transmission between routers e.g. through Request-Grant or threshold-based (on/off) link protocols, and end-to-end flow control dealing with packet scheduling and admission control, i.e. acknowledgment-based, credit-based, or threshold-based protocols [351].

In terms of link-level flow control, a flit usually moves only if the next buffer is not full. The alternative scheme where a flit may also move if the next buffer is full and a packet is simultaneously transmitted out of this buffer is much more complex to implement, since it requires significant back propagation time. In Spidergon STNoC the link level flow control is based on simplified credit mechanisms.

4.3   Communication Layering and Packet Structure

Similar to operating system and telecommunication protocols, global on-chip communication is decomposed into a stack of communication layers, ranging from physical to application software level. Similar to the ISO-OSI reference model, these layers interact through well-defined interfaces.

The OSI model is a hierarchical structure of seven layers that defines communication requirements among computing resources [351] [91] [37]. Within a particular layer, different entities are responsible for specific functions. All functions built within a layer can interact directly only with the layer immediately below, and provide a set of services to the layers above. Thus, changes within a protocol layer are prevented from affecting lower layers. The OSI model has successfully driven the definition and development of commercial local and wide area networks

As shown in Figure 4.5, the Spidergon STNoC IPU has been developed using a vertical layered approach based on a simplified version of the ISO-OSI reference model composed of only 4 layers. All these layers represent independent components that can be examined either separately or in relation to each other.

- The physical layer provides electrical media definitions that establish connection at the bit level among on-chip network resources. This layer determines the number and thickness of electrical wires for connecting together resources, and depending on technology establishes a type of synchronization. It also defines the number of repeater circuits to cope with long wires and corresponding control signals. On top of physi-

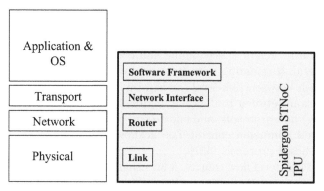

FIGURE 4.5:   Spidergon STNoC design

cal wires and circuits, the Spidergon STNoC link establishes a logical connection between on-chip network resources, through a robust data transmission protocol, through flow control [36] [35] [106] [380].

• The network layer is concerned with point-to-point packet routing from an arbitrary sender to an arbitrary receiver through the network topology. This layer implements wormhole routing and manages flit- and packet-level resource allocation and virtual channel and link arbitration.

• The transport layer hides network topology and link implementation issues, thus establishing efficient end-to-end flow control and source-to-target message communication services to higher layers, including operating system, system software, and application layers. In this layer, data are split or merged into NoC packets, and different processors or hardware blocks can be plugged as an open socket.

• Operating system, programmind models, system software, middleware and application software run on interface wrappers of higher NoC communication layers built upon the transport layer as abstractions of the lower-level communication framework. Thus, additional adaptation layers provide with standard or domain-specific communication and synchronization services, including support for efficient high-level programming paradigms.

With a vertical layered stack approach, decisions taken at different layers are not independent and interact with each other through well-defined interfaces. For example, the choice of the NoC topology influences simplicity of routing, chip area of the three hardware components, flexibility, and efficiency of communication services.

In Spidergon STNoC design a lot of attention has been placed on how the four architectural layers fit together and which type of issues must be addressed at each layer for designing an efficient and unique IPU. To simplify design, each layer of the stack is implemented by a specific component. Thus, Spidergon STNoC IPU consists of three hardware modules called link, router and network interface, and a software component (called software framework in Figure 4.5) that provides a set of communication primitives and low-level platform services.

This implementation implies that end-to-end protocols and services are orthogonal to topology routing and deadlock, and in turn these are orthogonal to physical link circuits. In fact, as far as we know, this is the first example of an IPU for Multicore SoC built as an ensemble of hardware and software components.

As discussed in Section 4.1, programming models are built on top of communication primitives. These primitives are implemented in Spidergon STNoC in a collaborative way by router, links, network interfaces and software framework components.

In particular, network interface plays an important role for realizing programming models since it implements communication primitives realized as "network transactions" [85]. A network transaction is a unidirectional transfer of data by a sequence of packets from an output buffer in the initiator to an input buffer at the target. When data are completely stored at the target, an appropriate event depending on the supported programming model is raised, e.g. interrupt service routine, process wake up, or access to external memory. This event is not visible by the initiator, who is usually notified through a reply message.

As discussed in Section 4.1, in case of shared memory programming models, communication operations and synchronization functions are mapped directly on communication primitives which are implemented in hardware by unified or split transaction protocols. Unified protocols, such as AMBA AHB, waste bandwidth, since the bus is locked from the request until the response phase. In this case, split transaction protocols, such as AMBA AXI, STBus, OCP, or VSIA, are implemented through two independent transactions: load or store. More specifically, a load is a request for transfering data from target to initiator, while a store is a request for transfering data from an initiator to a target. Each load/store transaction is implemented in two phases. In the request phase, one or more addresses and possible data are sent to the target, while in the response phase either data or acknowledgment is received by the initiator. For example, a load request that provides one or more addresses, and a load reply that receives the data. At the end, both initiator and target are aware of transaction completion. To resolve possible reordering occurring at the network, the network interface, or even the computing resources, each transaction is tagged with an identifier, so that the initiator is able to associate each reply with the right request.

Spidergon STNoC supports shared memory programming models through simple communication primitives that are implemented by split transaction protocols. Load or store network transactions are split into request and reply packets, but unlike existing bus and on-chip communication architectures, point-to-point network transfer is decoupled from protocol semantic. Network packets form an envelope that encapsulates all information required by split transaction protocols, but the way to move network packets from a source to a destination is agnostic of the communication primitive implementation.

Thus, interoperability among several different split transaction protocols is easily supported in Spidergon STNoC IPU. Although, currently only two protocols (AMBA AXI and STBuS) are fully supported, in the near future this list will grow to include other protocols based on shared memory programming models or other different communication primitives.

As described above, mapping of read/write operations in the shared memory programming model is straighfoward. For example, if a user process places a memory access, then the shared virtual address is translated to a physical one and a check is performed to see if the data are on the cache; if not, a load transaction is transferred to the right target through a load request phase. The target returns the data to the initiator during the load reply phase which completes the memory access. With more complex programming models, transactions composed of several packets with different semantics may be necessary to support specific communication primitives. For example, this is the case for cache coherence protocols.

In audio and/or video streaming applications, a message passing programming model is classically implemented on top of shared memory through store transactions. This model is usually based on split transaction protocols. A network node is a producer if it initiates the transaction and wants to transfer the data; while it is a consumer if it reacts to the transaction by consuming data. According to the message passing paradigm, a node can act as producer and consumer if it consumes and also produces data. Message passing support is enabled by Spidergon STNoC through a special communication primitive (called push) that supports actual inter-task communications. The push primitive can be easily implemented by a rather thin abstraction layer over native NoC network transactions. Push dynamically creates a type of virtual circuits among network nodes acting as producers and consumers, allowing direct transfer of streams with implicit control on buffer overflow. More details on the implementation of push are provided in Section 4.6.

Since interrupts for implementing task wake-up are slow, efficiency of inter-processor communication depends on how network transactions are interpreted. For this reason, Spidergon STNoC supports inter-process communication services using an ultra-lightweight version of Active Messages [372]. Active Messages are based on split (request and response) transactions combined with the possibility to associate a small amount of computation through an appropriate handler upon reception of a network transaction, i.e. a type of simple remote invocation. The handler type is carried by the network transaction and is executed immediately upon arrival. It is is similar to an interrupt, but more efficient, since it eliminates interpretation overhead.

Within this book, we only focus on communication services and features provided to support a shared memory programming paradigm. Other communication services are the subject of future work.

4.3.1   The Packet Format

A packet is a formatted data structure carried by the on-chip network based on the communication protocol rules. A NoC packet usually consists of control and data fields. Control information consists of header, possible checksum, and trailer. The data field is also called payload.

The header field provides all the necessary information to deliver user data to the right destination. In general, it includes the destination address(es), and sometimes also the source address. For variable size packets and depending on the link protocol it is usually convenient to represent the data length field first in the header field. The trailer field consists of a packet termination flag used as an alternative to a packet length sub-field for variable size packets.

A packet may include encoded priority information, burst identification, multicast addresses, or routing path selection, such as a virtual channel number, or a bit pattern identifying action at the next router. Moreover, the header provides an operation code that may distinguish request from reply packets, synchronization instructions, blocking or nonblocking load/store communication primitives, and normal execution from system setup or test. Sometimes performance related information, such as transaction identity/type, or flags for distinguishing and synchronizing access to buffer pools, e.g. for pipelining sequences of nonblocking operations, are included. In addition, if packets do not reach their destinations in their original order, a sequence number (and possibly a packet identification number) may be provided to help during packet reordering.

The checksum decodes header information (and sometimes data) for error detection or correction during packet transmission. In the presence of common errors, more complicated link-level credit-based protocols, such as alternating bit or go-back-n, may be implemented.

Finally, the data field is a sequence of bits that are usually meaningless for the channel. A notable exception is when data reduction is performed, e.g. in a combining, counting, or load balancing network [129].

Basic packets in simple point-to-point channels may contain only data. For complicated network protocols, e.g. supporting remote memory read/write operations, synchronization, or cache access, packets must use more fields, i.e. header, address, data and/or CRC.

The Spidergon STNoC network interface is defined so that multiple split transaction bus protocols could be encapsulated by the same network layer. Thus, as shown in Figure 4.6, the network layer payload consists of the overall transport layer packet, i.e. its header together with the payload. The network layer packet header contains an end-to-end QoS field and routing information. More specifically, routing information encodes a complete end-to-end path to arrive to the destination. In Spidergon STNoC, routing information is based

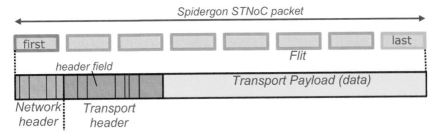

FIGURE 4.6:   The STNoC packet at network layer

on a destination address identifier and directional information. Exploiting the topological properties of Spidergon, the address identifier is 8 bits and the directional information is encoded in only 2 bits, independent of the number of nodes.

According to wormhole switching, Spidergon STNoC packet is split into one or more flits. The flit size is selected according to bandwidth requirements in the range of 16 to 512 bits; notice that flit sizes between 32 and 128 bits are the most appropriate for mapping requirements of most existing architectures. Depending on the packet and flit size, the network layer packet decomposed into a variable number of flits, each characterized by a tag specifying if the flit is at the beginning, middle, or end of a packet.

4.4   Routing Algorithms

The routing algorithm defines the path (or route) selected to transfer a packet from source to destination(s). A good routing algorithm balances load across different network channels, even in the presence of non-uniform and heavy traffic patterns. A well-designed routing algorithm also keeps path lengths as short as possible, thus reducing total packet latency. In addition, since routing must be fast and simple to implement, packet routing decisions at each switch must use locally available information only and not use any preprocessing.

Routing algorithms are defined depending on the way they select routing paths between source and destination nodes. Deterministic schemes always select the same path, even though multiple paths exist.

In oblivious routing, the path is selected without taking into account the present state of the network, i.e. other existing paths. Oblivious routing algorithms are classified into deterministic and randomized schemes. In deterministic oblivious routing, packets are sent along a more or less cleverly chosen path which only depends on the (source, destination) pair, while in *randomized oblivious* schemes, the route of any packet is independently chosen according to a probability distribution (which is a function of its origin and destination).

Valiant's famous two-phase routing algorithm is an example of a randomized oblivious scheme [364]. The two-phase algorithm consists of randomization (Phase I) and deterministic routing (Phase II). In Phase I, packets are sent to independent, randomly selected nodes through the network. The role of this phase is to reduce the difference between the average and the worst case performance. In Phase II, dimensions are corrected in an orderly [364] or random fashion [365], and packets follow a shortest path to their final destinations.

A common deterministic oblivious routing strategy for many symmetric topologies, such as grids of various dimensions and hypercubes, is the so-called dimension-order routing. In this method, a packet is routed along dimension 0 until it has reached its correct position along this dimension, then along dimension 1, and so on. Notice that the communication path is completely determined (a priori) by the source and destination address. Such deterministic routing strategies are non-adaptive; fixed and insensitive to network conditions, and thus they suffer from network hot spots and are also vulnerable to faults [361].

Therefore, several adaptive strategies, including randomized or deflection routing, have been proposed for improved performance under heavy traffic. While in oblivious deterministic routing, the path is uniquely defined by the source and destination address, in adaptive routing, the path is a function of the network traffic, i.e. routing events taking place within the network during its path, thus each router that receives a packet decides in which direction to send it [104].

Adaptive routing can be further divided into partially or fully adaptive. Since fully adaptive routing has a prohibitive implementation cost due to requirements for very large buffers, lookup tables and complex shortest-path algorithms, partially adaptive routing uses only a subset of the available physical paths between source and target. With local adaptive routing, for each move, packet transitions at any node depend only on the current node and destination, but they are independent of what happens in other nodes.

Randomized (or probabilistic) routing uses a sequence of random bits for making routing decisions. The routing algorithm utilizes randomization to convert the input distribution into a predictable or uniform one. Randomization is very effective when the deterministic algorithm suffers from unexpected worst-case instances, running on average much slower than in the random case. This situation is common for instances of highly structured communication and computation problems. Brassard and Bratley defined this type of randomization as Sherwood type [49]. Ben-David et al. conjectured that every randomized algorithm has the same performance as an appropriately designed online adaptive strategy [31].

Hot potato routing has been initially proposed for optical networks, since optical delay lines are expensive, and opto-electronic conversion for electronic buffering is slow [242]. Similar to circuit switching, hot potato routing (also called deflection routing) is an interesting switching strategy that does not require buffers at intermediate nodes [29]. With this technique, a packet is sent to the next network node, even before compiling its address information [170] [171]. Thus, packets continuously move, towards possibly a wrong direction, until they ultimately find a correct exit channel and reach their destination; proper deflection rules are needed to avoid deadlock and livelock problems. These rules restrict paths to non-minimal, and thus, performance of hot potato routing is not as good as other wormhole routing approaches [104]. Hot potato routing usually requires the assumption that each switch has an equal number of input and output channels. Due to its simplicity, this scheme has been proposed in the NoC domain (see Nostrum in Section 1.1). However, it does not seem applicable for an effective NoC implementation, due to re-ordering problems and unpredictable control of the packet latency in high load conditions.

An alternative approach to probabilistic or adaptive routing which minimizes contention of packets directed to memory, especially for irregular communication patterns, involves random polynomial hashing of the distributed shared memory space, i.e. storing data at random modules in the shared memory. Although this approach (if used on individual words instead of groups, such as cache rows) destroys spatial data locality, since applications have no control over data placement, it makes typical routing problems appear as random. Depending on the application, the locality problem can be resolved by moving, copying or caching data leading to hot spots or by using a combining network. A combining network is a switching system for accessing memory modules in a multiprocessor, in which each switch remembers the memory addresses used, and can then satisfy several simultaneous requests with a single memory access. The approach of randomly hashing shared memory dramatically reduces network congestion and memory contention, since requests are randomly distributed throughout network links and memory modules, and the probability of mapping many keys to the same index is low. This technique has been used in the Tera multithreaded supercomputer, the IBM RP3 which implemented a combining network, and the German hardware PRAM prototype [2] [257]. Although in general all above techniques are attractive, they are not effective within the NoC context. The main reason is high implementation complexity and packet reordering problems.

4.4.1   Spidergon STNoC Routing Algorithms

Spidergon STNoC adopts oblivious routing. Several alternative routing algorithms are proposed, so that the application user can select the best solution for a specific network topology instance and application traffic scenario.

The communication path is completely determined by the source and destination address and the packet route is fixed when packets enter in the network. The choice for deterministic routing guarantees ordered end-to-end communication at network level, hence avoiding costly flit reordering at packet reception. Out-of-order communication is possible as a protocol feature at the transport layer. For example, responses can return in different order with respect to requests when accessing different slaves, or when a slave implements some reordering policy for advanced QoS-based memory scheduling.

Spidergon STNoC routing schemes are simple since they leverage the symmetry and simplicity of the Spidergon topology. Consequently, the relevant implementation of the on-chip routers is extremely efficient without requiring expensive routing tables. Spidergon STNoC routing information is encoded in the packet header at network injection points or network interfaces. Thus, routers can easily decode the path from the header. This type of routing is known as source-based routing, since the path is completely determined at packet generation time and is fully encoded in the packet header as routing information.

Although attractive for its great simplicity, a drawback of standard source-based routing is a variable network header size. In general, if the direction after each network hop is specified separately, then this field would increase with the network size; in addition, the router should be provided with a shift mechanism to access the correct part of this routing field. However, for Spidergon STNoC, routing information has a fixed length of a few bits due to exploiting regularity and sense of direction to encode shortest paths. This decreases routing logic complexity and allows for a reduced router pipeline depth.

Although Spidergon STNoC routing is not adaptive, i.e. selection of the routing path is fixed upon packet injection at the network boundary, it is programmable by supporting efficient software control of packet routes. A routing function executed at packet injection time associates a particular network path to the packet depending on the source and destination address. This function can be changed during runtime through software reconfiguration, fully exploiting topological path redundancy.

A primary goal of routing programmability is fault tolerance support which is foreseen mandatory in submicron technology. There are many other important aspects, which are enabled by this feature. Smart traffic management for different application use cases can be based on network path allocation. This type of coarse grain adaptiveness redistributs interconnect load to limit network congestion, improving QoS. This is especially important for managing on-chip interconnects with complex memory hierarchies in which increased concurrency relies on selecting better network paths to accommodate traffic in different ways. Moreover, this feature can be exposed through the Spidergon STNoC software framework or integrated within OS kernel memory management, application task embedding, or even at user-level, within the application itself.

Another crucial motivation in exploiting software routing reconfiguration is productivity. The on-chip communication infrastucture can be designed quickly, allowing routing to be refined in a late stage of product life. In this way, potential bugs, conflicts, unpredictable software-dependent flows, new IPs, or wrong bandwidth estimations can be managed with flexibility through the Spidergon STNoC IPU.

Packet routing decisions are carried out using only local information available at each network node; each router has a unique address i in the network, $0 \leq i < N$, where N is the network size. Since the routing algorithm is local, and the Spidergon topology is vertex- and edge-transitive, the routing algorithm is equivalent at any node .

When a router receives the first flit, i.e. the network header of a new packet, the forwarding path is computed. The Spidergon STNoC routing algorithm compares the network address of the current router to that of the destination router, i.e. the router connected to the target resource; the latter address is encoded in the packet header. If the two network addresses match, flits are routed to the local port of the router. Otherwise, an attempt is made to forward the flit towards a clockwise or counter-clockwise direction along the ring, an across link, or in a hierarchical way to another instance of the Spidergon topology family.

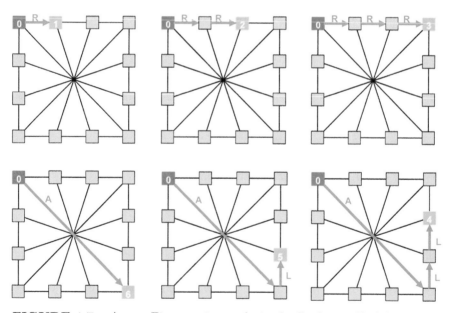

FIGURE 4.7:   Across-First routing paths in the Spidergon NoC for $N = 12$

The first algorithm, called Across-First, moves packets along the ring, in the proper direction, to reach nodes which are closer to the source node, and use the across link only once at the beginning for destinations that are far away. As an example, Figure 4.7 shows routing paths with the Across-

First deterministic shortest-path Spidergon routing algorithm on a 12-node Spidergon STNoC. The clockwise direction is indicated with R (right), the counter-clockwise is L (left) and the across link connection is A (across). Since the Spidergon topology is vertex symmetric and the routing algorithm is local, i.e. symmetric for all nodes, without loss of generality, Figure 4.7 shows routing from node 0 to half of all possible destination nodes.

The Across-First routing relies on a simplified form of source-based routing, since the entire path can be encoded in the packet header using the following two observations. First, the Across communication link is selected at most once, always at the beginning of each packet route. Second, when the packet starts moving along the ring, it follows the same direction (right or left) for the entire path. According to these properties, only packets arriving from the local resource or from the across port need to be considered for routing. All other packets continue to move in the same port until they reach their destination, i.e. clockwise or counter-clockwise.

These properties are exploited by Across-First routing strategy using two separate routing phases. At first, there is an initial pre-computation during packet generation at the NI, and then an extremely simple and fast source-based routing phase performed at the input ports of each router. As already stated, this second phase requires no computation or lookup table, unlike complex routing algorithms for other smaller diameter, circulant, chordal, or double loop graphs. Across-First routing in Spidergon requires a fixed size of directional information (2 bits) that specifies the packet's path through the network: right, left or exploiting the across direction. For instance, in Figure 4.7, when the destination is 5, the first hop uses the across link and the direction field is set to "across-then-left" value, that indicates the cross as first hop and then left as ring direction. When destination is 6, direction is still "across-then-left", but left is not used since the packet arrives at destination before moving on the ring. Notice that during packet generation time, source and destination fields in the packet header are set using the addresses of the source and destination nodes. In this phase, node destination is decoded from higher level protocol information. Depending on the adopted transport layer, specific translation functions are implemented at the network interface to map the protocol address into a network address.

The following pseudo-code in Figure 4.8 describes the first step of Across-First routing scheme. This step can be implemented by using either dedicated logic or lookup tables.

After performing the initial routing step during packet generation at the network injection point, a very simple and fast algorithm can be used to forward the packet at each router, exploiting the destination and direction fields of the header.

Assuming that destination is the network address of the source node, and current is the address of the node crossed by the packet, Figure 4.9 indicates the pseudo-code for the routing function during each step.

The Across-First algorithm is appropriate not only for efficient one-to-one,

```
//initial step at packet injection (network size N, diameter d)
if (destination - source) mod N in {1, 2, 3 …. d}
    first hop in clockwise (Right) direction;
    set the direction field to Right;
else if (destination - source) mod N in {N-1, N-2, N-3 …. N-d}
    first hop in counter-clockwise (Left) direction;
    set the direction field to Left;
else if (destination - source) mod N in {d+1, d+2, …. N-d-1}
    first hop in across (Across) direction;
    if (destination - source) mod N in {d+1, d+2, …. N/2}
      set the direction field to Across-then-Left;
    else
      set the direction field to Across-then-Right;
```

FIGURE 4.8: Pseudo-code for initial step of Across-First routing

```
//second step at each packet router
if (current = destination)
  route packet to the local network interface direction
else
  route packet in the direction specified by direction field;
```

FIGURE 4.9: Pseudo-code for routing step of Across-First routing

but also for one-to-many and many-to-many traffic patterns. However, it is not optimal for many-to-one and related few-to-many traffic, i.e. when many initiators communicate with few targets.

Hence, similar to Across-First, another routing scheme can be used on the Spidergon STNoC. Instead of jumping through the cross link as first hop and moving along the ring to reach the final destination, packets can first move along the clockwise or counter-clockwise directions and finally take the cross link to the destination node. Due to the use of cross links as a last hop, this routing scheme is called Across-Last. Alike Across-First, two routing steps (initial and routing) can be identified, while the pseudo-code is also similar; the algorithm actually requires only a simple modification in the initial phase of Across-First, i.e. modifying the distance computation, or the destination address information in the packet header. An example of this scheme is shown in Figure 4.10.

Both presented routing algorithms are shortest-path. Moreover, as discussed later in Section 4.5.1.1, Spidergon STNoC also allows routing in many other ways. For instance, packets may progress along the ring without using the cross connections. In this case, the path is not optimal, but it could help distribute the load of request/response flows depending on the specific application requirements.

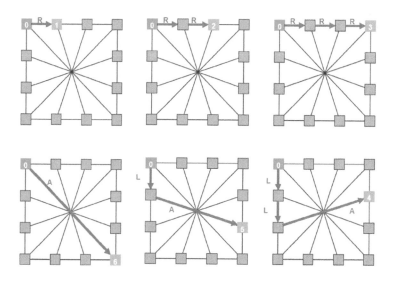

FIGURE 4.10: Across-Last routing paths in the Spidergon NoC for $N = 12$

4.5 Livelock, Starvation, and Deadlock

To ensure correct functionality, the NoC must always be free of livelock, starvation, and deadlock.

For non-FIFO buffers and adaptive routing schemes, potential livelock and starvation problems must be solved. Livelock refers to packets either indefinitely waiting at a network buffer due to an unfair queuing policy, or circulating the network without making any progress towards their destination. Starvation refers to packets indefinitely waiting at a network interface buffer due to an unfair arbitration policy. Notice that both livelock and starvation reflect problems of fairness in network routing or scheduling policies. Livelock usually does not occur with shortest path or randomized routing and can be avoided with adaptive (usually non-shortest path) routing strategies by implementing age-based priority schemes.

For all routing schemes presented in this book, Spidergon STNoC is always free of livelock and starvation. These issues are addressed through fair scheduling policies of shared resources.

4.5.1 Low-level Deadlock

A critical issue in NoC design is that the network must guarantee deadlock-free operation. Deadlock occurs when a group of packets are unable to progress, since they are mutually blocked waiting on each other to release shared resources, buffers or virtual channels that are never let free. A low-level deadlock is caused by a cyclic dependency of not granted packet requests for shared interconnect (buffer or channel) resources at the network layer.

Deadlock is a cataclysmic network event, since blocked chains of buffers due to deadlocked packets easily cause an avalanche of similar effects paralyzing network operation. This is especially true for wormhole switching in which flit-level flow control causes additional deadlocks. In order to prevent deadlock, two solutions are possible: a priori deadlock avoidance or recovery from deadlock.

In case of oblivious packet routing, deadlock can be avoided by eliminating cycles in the channel dependency graph (CDG). This condition is only necessary but not sufficient in the case of adaptive routing [104], i.e. it is possible to have oblivious routing algorithms with cyclic channel dependencies while still being deadlock-free; however, notice that current examples of this behavior require unusual routing function and topologies. The CDG is a directed graph where the vertices are the network topology channels and the edges represent the routing algorithm. More specifically, if a routing function allows packets arriving at node i over channel a to leave i over channel b, then the CDG will have an edge between a and b, directed towards b. The CDG depends on the network topology and the associated routing algorithm.

The assumption behind use of the CDG is that network extraction points act as sinks, i.e. they are not involved in potential dependency cycles. This hypothesis is related to the fact that packet semantics are not considered at the network layer. When transport layer protocols are involved, network end-points can introduce dangerous dependencies. If deadlock is caused by dependencies external to the network layer, this is called high-level deadlock or protocol deadlock (see Section 4.5.2). For instance, a simple request/response protocol can lead to deadlock when dependencies occur in target devices between incoming requests and outgoing responses.

Deadlock avoidance can be achieved either by construction through deadlock-free routing, or using virtualization of paths which enables packet moves on ordered channels, as explained in the example below. Upon deadlock detection through additional hardware, deadlock recovery techniques implement complex restoring mechanisms which are a type of functional reset of the network. In fact, not only restoring is complex, but also distributed detection requires important additional hardware costs.

Figure 4.11 shows a unidirectional ring with a trivial routing, since a single path is possible between any couple of nodes. The right hand side of this Figure indicates the corresponding CDG, which clearly exhibits a cycle.

A possible deadlock condition occurs when a situation similar to Figure

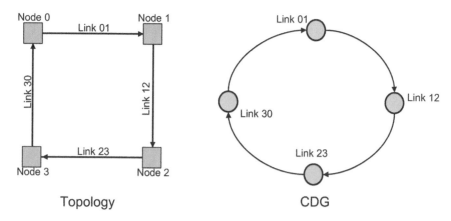

FIGURE 4.11: A simple unidirectional ring with corresponding CDG

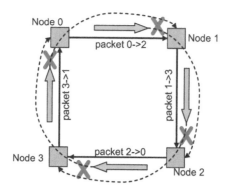

FIGURE 4.12: A deadlock condition in the 4-node unidirectional ring

4.12 occurs [86]. A packet from node 0 to node 2 is waiting blocked at node 1. The link between node 1 and 2 is blocked by another packet from node 1 to node 3 which is waiting blocked at 2. Again another packet from node 2 to node 0 is waiting blocked at node 2. Finally, at node 3 yet another packet moving from node 3 to node 1 is blocked waiting for the link from node 0 to 1 which is blocked by the first packet moving from node 0 to node 2. This cycle of dependencies causes an infinite stall.

One way to provide deadlock-free routing is by using virtual channels (VCs) to break cycles in the CDG. This technique provides logical links over the physical channels connecting two ports and establishes a number of independently allocated flit buffers in the corresponding transmitter/receiver nodes (each VC is associated to one buffer).

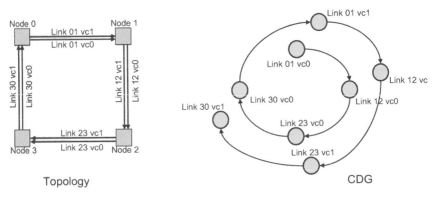

FIGURE 4.13:   Use of virtual channel for deadlock avoidance

Figure 4.13 shows the 4-node unidirectional ring with virtual channels indicated as separated arrows; the physical link is unique. The CDG is actually a virtual channel dependency graph, since the shared resources to be represented are the independent buffers assigned to each virtual channel. According to CDG definition, each node on the right hand side of Figure 4.13 is a separate virtual channel, hence this graph has 8 nodes. The routing algorithm must define not only the way packets move in the network, but also the virtual channel in the CDG to use. For deadlock-free routing, a packet has to use virtual channel 1 when the source node i must send a packet to a destination node j and j is greater than i, otherwise virtual channel 0 is used (if j is less than i). For example, assume that node 1 must send a packet to node 3. Since 1 is less than 3, node 1 uses the channel "link12.vc1" to first send the packet to node 2. Node 2 compares its network address to its destination (3), and chooses channel "link23.vc1". If node 1 has to send a packet to node 0, it chooses channel "link12.vc0" to send the packet first to node 2. At node 2 again the selected channel is channel "link23.vc1". Finally, at node 3 the packet directed to node 0 moves through channel "link30.vc0". A packet from node 3 to node 1 will travel on "link30.vc0" as a first step, and "link01.vc1" as a second step. The most important result of applying this rule is that it creates a total order of the channels, or in an equivalent way the CDG does not contain a cycle anymore, as shown in the right side of Figure 4.13.

4.5.1.1 Deadlock Avoidance in Spidergon STNoC

This Section focuses on the Across-First scheme, although similar implications based on an identical CDG can be drawn for the Across-Last routing.

Spidergon STNoC with Across-First routing scheme is not deadlock free due to cycles in the CDG, as shown in Figure 4.14. Notice that this CDG is attributed to a Spidergon topology of generic size. Spidergon STNoC behaves essentially as a bidirectional ring, with across connections used only as first hop. Both CDG cycles arise from dependencies in the bidirectional ring, while cross links are not part of cycles since they are not intermediate steps in the packet route.

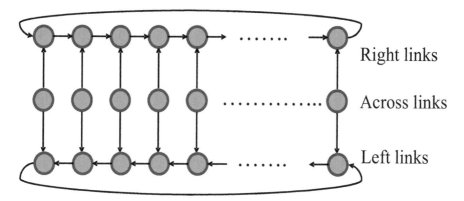

FIGURE 4.14:   CDG for Across-First routing on Spidergon NoC topology

Hence, low-level deadlock avoidance in Spidergon STNoC is analogous to that in the ring topology. Dependencies in the ring can be removed using two virtual channels in the clockwise (R) and counter-clockwise (L) ring direction [87]. Consequently, we extend the Across-First routing method presented in Section 4.4.1 to the following naïve deadlock-free virtual channel allocation algorithm.

The first virtual channel is always used, if and only if the network address of the current node is greater than that of the destination; otherwise, the second virtual channel is selected. This naïve algorithm uses a physical dateline at node 0, and provides no virtual channel balance. More specifically, any packet traveling along the clockwise (or anti-clockwise) communication link of the Spidergon ring initially uses virtual channel 0, and upon crossing the dateline node it uses virtual channel 1. Since buffer dependencies cannot cross the dateline, no circular dependency exists along the ring.

Optimized "load-balanced" virtual channel allocation on Spidergon may provide efficient use of network buffer space, thus improving performance by reducing network contention. On the contrary for the naïve allocation, many buffers associated to virtual channel 1 are unused. Virtual channel 1 can be used for at most the $N/4$ nodes after the dateline; this is the maximum distance a packet may travel on the Spidergon ring after the dateline node. Physical dateline assignment for the naïve algorithm leaves considerable flexibility for a more balanced virtual channel allocation. For example, if a packet route does not cross the dateline, any virtual channel can be used, i.e. virtual channel assignment for these packets is unconstrained. Thus, logical datelines can be exploited further.

Virtual channel allocation based on proper assignment of unconstrained packet routes can avoid deadlock conditions and exploit fairness to increase system performance.

A simple origin-based virtual channel allocation algorithm has been developed to appropriately combine deadloclk avoidance and load balance on Spidergon STNoC. The origin virtual channel allocation algorithm is based on partitioning the Spidergon ring into two contiguous halves: nodes $\{0, N-1, N-2, \ldots N/2+1\}$ are called type I, while nodes $\{N/2, N/2-1, \ldots, 1\}$ are type II nodes. As already stated, with shortest-path deterministic routing, the maximum distance of an Across-First path on the ring is $N/4$. Thus, no packet can move out of one partition, into another, and return to the first one. With origin-based virtual channel allocation, packets originating at type I or type II nodes that do not cross the two dateline nodes ($N/2$ and 0) are assigned to any of the two virtual channels, i.e. either virtual channel 0 or virtual channel 1. However, in order to avoid deadlock, packets crossing the dateline nodes, i.e. packets originating at type I nodes crossing node $N/2$, or packets originating at type II nodes crossing node 0, are initially routed on virtual channel 0, and then upon crossing their dateline they are shifted to virtual channel 1. Notice that packet (flit) reordering for packets originating at the same node and heading to the same node is still avoided through a static assignment of virtual channels.

By assuming random traffic, i.e. random destinations or permutations, multinode broadcast, i.e. one-to-many relations, or total exchange traffic patterns, each communication link will carry approximately the same number of packets, thus leading to better utilization of virtual channel buffers, proportional to the number of packets that pass through it.

By examining the relative probability of a packet crossing a dateline, a metric of fairness for the origin-based allocation scheme can be computed. The virtual channel assignment is fair on $2(N/4 - 1)$ nodes with packets equally assigned on both virtual channels. It is in favor of virtual channel 0 by 2:1 on $N/4$ nodes, in favor of virtual channel 1 by 2:1 on $N/4$ nodes, and completely unfair on only two nodes (those required for deadlock-free operation). This analysis is also true for all-to-all patterns, while with locality of references and most latency hiding models even better performance is expected, since most packets become unconstrained as they do not cross system datelines.

The presented origin-based allocation improves resource utilization compared to naïve allocation, while it is still very simple to implement. Moreover, this model can be extended to a more general setting. For example, Figure 4.15 shows a fine grain renumbering scheme which provides statistically better virtual channel balancing by breaking the CDG at four points for random traffic.

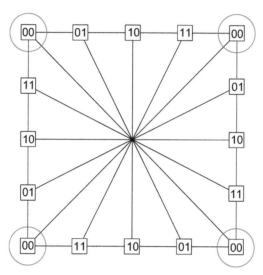

FIGURE 4.15: Spidergon network for $N = 16$ with destID numbering

As discussed before, Across-First and Across-Last routing avoid routing-dependent deadlock on the STNoC Spidergon by utilizing a pair of virtual channel resources only in the clockwise and counter-clockwise physical links and no virtual channels in the across physical links.

Another way to avoid deadlock is to design routing schemes that are free of dependency cycles by construction; this is normally achieved by restricting the possible paths in the topology. The set of Spidergon STNoC routing schemes can be extended with deadlock-free routing algorithms by construction.

Both the Across-First and Across-Last allow simple, low-cost variants, called Zero, without virtual channels. Zero variant routing algorithms use a single virtual channel in all directions and avoid deadlocks by limiting network paths that travel on the ring and cross node 0 (or any other given node, due to network symmetry), essentially utilizing the Spidergon ring as a linear array.

Thus, for both algorithms, only the initial phase is modified to let routes of packets that cross node 0 take an alternative route, which no longer forms a shortest path, but is short due to using cross link. Moreover, many Multicore SoC application tasks can be mapped efficiently onto the Spidergon topologies without requiring any paths to cross a particular node. Then, there is no penalty involved with using zero variant routing, thus fully exploiting reduced router complexity for a low cost virtual channel-free solution.

Virtual channels not used for low-level deadlock avoidance can be used for other sophisticated mechanisms related to network virtualization, security, performance, and QoS.

4.5.2   Protocol Deadlock

As discussed in Section 4.5, deadlocks in NoCs can be broadly categorized into routing-dependent deadlock that occurs within the NoC and message-dependent or protocol deadlock that occurs at the network interface [248] [326].

Actually, most current NoC designs are free of routing-deadlock. However, they may still be susceptible to deadlock due to message-dependent (or protocol) deadlocks due to dependencies at network end points (NI) [192].

Protocol (or high-level) deadlock occurs when interactions and dependencies are created between different message types at network endpoints, and these messages share resources in the network. Even if a network is designed to be routing-dependent deadlock free, protocol deadlock can block the network indefinitely due to finite network buffer size or end-to-end communication resource dependencies between incoming requests and outgoing responses, affecting system operation and reliability.

In Section 4.5.1, the CDG shows potential deadlock cycles, assuming that end points act as sinks, i.e. they are (sooner or later) able to extract packets from the network. It is the removal of this hypothesis that leads to potential protocol deadlock, since injected packets may actually share buffer resources or exhibit protocol dependencies with extracted packets.

While research on routing-dependent deadlocks is widely documented, relatively few general results exist on avoiding or recovering from protocol deadlocks.

In case of split transaction protocols implemented on top of NoC packets switching, protocol deadlock may arise due to finite network buffering (lack of resources) or circular dependencies between packet requests and replies. For example, a circular dependence between two processors accessing each other's caches can arise if each cache waits for its reply before servicing its requests, and the buffers between the two processors are completely full of unrelated packets in both directions. Buffering and handling of packets within a hierarchy of complex write-back caches in shared memory architectures introduce a deep level of dependencies which is a potential source of protocol deadlock [128].

Next, we discuss common avoidance and recovery solutions to protocol deadlock. Avoidance methods are broadly categorized into buffer dimensioning, end-to-end flow control, virtual channels and strict ordering; while a well studied recovery technique is two-case delivery.

4.5.2.1 Buffer Dimensioning

A simple theoretical remedy for protocol deadlock among packets that experience resource conflicts is to provide additional resources. Since buffer memory is usually the limited resource, buffer dimensioning provides sufficient amount of buffer space at network endpoints to ensure that there never will be congestion. With this approach, all buffers at network nodes are considered infinite, or enough to hold the maximum possible number of outstanding transaction requests. This method can be implemented through either distributed memory lists or input buffer space, e.g. by augmenting hardware queues in software, as in MIT Alewife [192]. Unfortunately, in the Multicore SoC the global interconnect area has to be minimized.

4.5.2.2 End-to-End Flow Control

End-to-end flow control mechanisms deal with individual flows between producer and consumer components, i.e. ensuring that the producer does not send more data than the consumer can accept and vice-versa. Thus, these mechanisms guarantee enough buffer space for storing the packet at its destination to prevent message-dependent deadlock [119].

Similar to congestion control techniques, end-to-end flow control methods based on deflection routing, packet drops, or return to sender do not require resource reservations.

End-to-end flow control methods that require resource reservations include sliding windows techniques or credit-based end-to-end flow control. In the latter case, a node is allowed to send a packet if and only if sufficient space is available at the destination node; notice that for efficiency reasons, credits may be piggy backed within packets traveling in the opposite direction; however, notice that this solution is not general enough, since it relies on customized protocols.

4.5.2.3 Congestion Control

Notice that in contrast to end-to-end flow control, congestion control regulates traffic flow in network buffers and links to avoid network overload and reduce resource contention in time and space. Thus, congestion control effectively improves global quality of service.

Several congestion control methods from computer networks can also be applied to NoC. For example, admission control techniques based on time-division multiplexing, virtual circuits, and traffic shaping using leaky or token bucket schemes use average- or worst-case resource reservations to reduce or even eliminate contention. In contrast, techniques without reservation, such as adaptive or deflection routing, and packet drop reduce the network injection rate to improve quality, but cannot provide hard guarantees on end-to-end bandwidth or latency.

4.5.2.4   Strict Ordering and Duplication of Physical or Virtual Networks

Strict ordering is a common way to avoid protocol deadlock in multiprocessors. It orders network resources by grouping messages which do not have any logical (protocol-related) conflicts into a physical (or virtual) channel. An extreme type of strict ordering is in virtual networks, where each message type has its own separate logical network. Thus, resource conflicts occur between messages within a virtual channel, while logical conflicts occur between messages in separate virtual channels. Since messages which experience resource conflicts do not experience logical ones, and messages which experience logical conflicts do not experience resource conflicts, deadlock-free operation is achieved by breaking cycles created from resource sharing by different messages.

This practical solution separates request and reply networks, as either two distinct physical or two logical networks multiplexing physical channels over separate virtual buffers, and gives priority to reply messages. This solution is sufficient for strict request-reply protocols, as in SGI Origin 2000 or Active Messages [374].

In a strict request-reply protocol, a request generates a reply (or nothing), while a reply does not generate any other transactions. In this case, to avoid deadlock, an incoming request can be serviced only if there is space in the outgoing network interface reply queue, while an incoming reply may always be serviced since it does not generate any further packets. However, most latency-hiding protocols in shared memory multiprocessors, including cache coherence protocols, are not strictly request-reply type; for example, an exclusive read may generate invalidation requests, which in turn generate acknowledgment replies. For such general protocols, as many virtual networks as the longest chain of different transaction types are needed, so that the final transactions (those that do not generate any other transactions) can always make progress. This approach is expensive due to duplication of virtual channel buffers, and also inefficient, since most channels remain largely underutilized.

The Stanford DASH multiprocessor coherence scheme is shown in Figure 4.16 [192]. According to this protocol, a node A directly requests data from another node B. Then, node B sends an invalidate message to the current owner of the data copy, who replies by sending a copy of the data to both nodes A and B. For this protocol, there are 4 message types and two types of message dependencies which correspond to the two cycles in Figure 4.16, requiring at least three independent virtual channels for deadlock-free operation [180].

Deadlock freedom is even harder to ensure in message passing-based Multi-

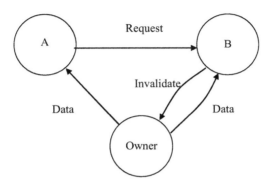

FIGURE 4.16: Cache coherence messages in DASH [192]

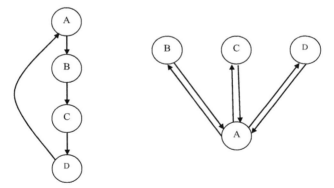

FIGURE 4.17: Alternative representation reducing the depth of MDG [192]

core, since user applications can dynamically create messages with a variable depth of dependency. In general, in order to analyze vulnerability to dead-locks, similar to a cyclic dependency graph, the message dependency graph (MDG) examines dependencies among different messages. In MDG, vertices represent terminal nodes, while edges model protocol messages. The number of virtual channels that are sufficient for message-dependent deadlock free network operation is equal to the depth of transactions in its MDG. There are ways to reduce a depth k MDG by splitting it into k transactions of depth 2 starting from the initiator and ending in every other node in the graph; for example, Figure 4.17 shows an MDG of depth 3 split into 3 graphs of depth 2. However, MDG depth reduction introduces additional protocol complex-ity and decreases protocol performance due to additional round trip delays and difficulty of the initiator to allocate enough space for all intermediate messages.

However, apparently at the expense of performance degradation, protocol dependencies can be converted to simple request-response dependencies which can be eliminated using either two disjoint physical networks with a separate physical data bus for requests and responses or two separate virtual networks that handle requests and responses.

4.5.2.5   Recovery Technique: Two-Case Delivery

The most common among past solutions to message-dependent (protocol) deadlock is utilization of virtual channels. However, virtual channels add significant complexity to the channel arbitration logic, and overall they do not improve average network throughput, especially with hot spot traffic. Most important the virtual channel technique is not cost effective when message exchange protocols are not of request-response type.

Deadlock recovery is an alternative approach based on a recovery phase that does not restrict the way that the routing algorithm employs the network ports, but rescues the program from deadlock, if one arises. Recovery does not have to be particularly fast, since deadlocks are usually rare.

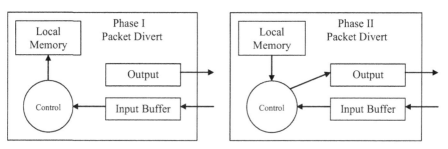

FIGURE 4.18:   Two-case delivery for protocol deadlock recovery [192]

A very powerful deadlock recovery technique, used in the Alewife multiprocessor, is two-case delivery [5]. This method internally emulates an infinite buffer space at the end nodes of the message transfer and preserves a strict request-reply protocol by detecting and recovering from a possible deadlock or data race. Figure 4.18 illustrates the phases of two-case delivery [192]. With this technique, whenever input queues are full, an interrupt notifies the core to store in its local memory all packets currently in the input queue and send negative acknowledgment messages to the requestors (divert mode of operation). This redicrection of data into the local memory occurs until the output request queue is no longer full or the request at the head of the queue does not generate replies. The empty space created at the queues allows previous nodes in the dependency graph to make forward progress. Then, when deadlock is not present anymore, packets stored in local memory are rescheduled by placing them in the output queue (relaunch mode). This assumes that the retry from the requester will eventually succeed. With extreme pathology, a deadlock, livelock or starvation may occur again during relaunch. Thus, alternative actions are sought; for example, to avoid livelock or stravation, the priority for request packets must be increased by the number of times a request has been negatively acknowledged.

A major issue for two-case delivery is correct detection of the occurrence of deadlock hazards. If only input queues are examined, then deadlock detection can incorrectly set node hardware to divert mode. For example, consider a processor which runs a database and an Apache server. If the database server is given higher priority than the Apache server, then the network interface is ignored for considerable time, and thus input queues may be full resulting in false deadlock interrupt. Similarly, examining only congestion in output queues close can result in false deadlock detection. Thus, the most reliable scheme for detecting deadlock utilizes information on a full and not moving output queue, and simultaneously a full and not moving input queue.

Except for the MIT Alewife system [5], two-case-delivery has also been implemented in the MIT Raw architecture [353]. Raw uses light processors with local SRAMs connected together and to off-chip memory through two separate physical static and dynamic 2-d mesh networks. While for the static network, the compiler guarantees deadlock-free fine grain parallel processing by defining packet routing at compile time, the dynamic network is totally exposed to two major groups of accesses: random or unknown user-defined communication protocols and unknown at compile time, memory accesses which can potentially deadlock. Thus, the dynamic network targets interrupts, I/O, DMA and other irregular off-chip communications [352]. Notice that protocol deadlocks in Raw always involve at least one packet in the dynamic network.

The dynamic network is based on two disjointed (request and reply) virtual networks: a high priority network for off-chip memory bank access (DRAMs) employing a simple deadlock avoidance scheme, and a low-priority network for communication among processor cores using deadlock recovery based on two-case delivery. Notice that for performance reasons, DRAM devices are located at separate sides of the 2-d mesh.

Deadlock detection in the low priority dynamic network slightly differs from the Alewife approach. A watchdog timer is incremented during each cycle. It is reset to zero upon a successful read to the dynamic network input port. If the counter reaches a predefined user-specified value, then an interrupt is fired, indicating potential protocol deadlock. This deadlock may be true or false positive, i.e. a node may consume a packet at a very slow rate, sufficient for the counter to reach a deadlock-critical threshold value. Then, network performance is further reduced, since the consumer processing element also executes the deadlock recovery code.

FUGU workstation architecture dealt with this problem in more detail by implementing network resources that can decide whether the potential deadlock is an actual one [222]. However, techniques that separate actual and false positive deadlock detections always require expensive deadlock detection algorithms and increase NoC implementation complexity.

Once deadlock is identified, a recovery phase must be initiated. During deadlock recovery packets blocking the network are drained in an SRAM connected to the processor core. Thus, valuable memory space is freed and packets are able to move again along the deadlocked path. The processor core in which deadlock recovery code has been executed must subsequently access through its network interface the SRAM so that first packets stored are first consumed. This solves out-of-order delivery problems, since packets stored in the SRAM are read and processed based on their order of arrival conventional execution resumes.

Extra care should be taken in defining a sufficient size for the SRAM. The number of outstanding network transfers should be finite and relatively small. However, in the worst case all processor cores may communicate with one processor core. In general, SRAM size depends on the number of outstanding network transfers and the specific communication schemes implemented by the user-defined communication protocols.

4.6   Protocol Deadlock Avoidance in Spidergon STNoC

Using virtual channels and the previously described routing algorithms, Spidergon STNoC IPU is free from low-level deadlock. As described in Section 4.2, the end-to-end protocols supported by Spidergon STNoC, in particular split transaction and message passing protocols, introduce dependencies between messages that could lead to potential high-level protocol deadlock.

Let us consider a transaction between two messages m_1 and m_2. The notation $m_1 \prec m_2$ represents a dependency chain between the two types of messages, where m_2 denotes the terminating message, i.e. it does not generate subordinate messages [325]. Then, with strict ordering or virtual networks, a NoC is deadlock-free if each message type has its own logical (physical or virtual) network.

As far as the split transaction protocol is concerned, a simple way to remove protocol deadlock is to have two separate physical (or virtual) channels, one for transferring request packets of the transaction and the other one for transferring reply packets.

The clean layering approach, implemented at all levels by Spidergon STNoC technology, provides great flexibility to the Multicore SoC architect to explore alternative scenarios. For example, in each physical or virtual network, any deadlock-free routing algorithm from Section 4.4 can be used; it is possible to use the zero variant of Across-Last routing in the request virtual network, and the zero variant of Across-First routing in the response virtual network, or vice-versa.

For the message passing protocol the situation is more complex. In contrast to split transaction protocols, with message passing protocols the message dependency chain is $m_1 \prec m_2 \prec ... \prec m_n$ since each message can generate another one. Each message type has to be routed through its own independent virtual network. However, the maximum message dependency length (n) is not always known a priori since it depends on the application. To dimension the NoC with a potential worst-case has extremely high cost, since resources necessary to support several virtual networks increase NoC complexity and they could also be not well utilized.

Among other avoidance techniques, buffer dimensioning is too costly and depends on the application. These limitations are no longer valid with end-to-end flow control, since no packets are ever injected in the network that cannot also be consumed. The most common end-to-end flow control is based on credit exchange between the consumer and the producer NI, alike the best effort scheme in Philips Æthereal NoC [96]. Credit-based end-to-end flow control schemes resolve message dependencies but require dedicated buffer space. Cost increases proportionately when multiple producers send data to the same consumer, unless buffers at the consumer NI are shared, as explained next in the solution used by Spidergon STNoC called "Connection Through Credit algorithm" (CTC).

CTC is one possible implementation of the push communication primitive introduced earlier. First, the producer opens a virtual connection towards the consumer through the network, sending the number of flits to transmit. Unlike circuit switching, network resources are not locked, hence several streams can be efficiently interleaved and NoC resource utilization is maximized. At consumer side the required shared buffer resource is reserved and the producer is informed by sending back an acknowledgement. Now, the producer can send actual data without conflicts from other potential producers. The data transfer over the established virtual connection is regulated by remote buffer overflow control which uses specific network packets sent from consumer to producer.

CTC protocol introduces a dependency chain of length three, since the push primitive is implemented with three message types. These messages do not introduce any deadlock, since the dependency is broken at different levels. Messages that are short are guaranteed to sink upon arrival due to proper buffer dimensioning, while long data messages are injected with a guarantee by protocol construction that they will not overflow at the receiving buffer. Thus, using the arguments in [203], it can be shown that the CTC protocol guarantees freedom from protocol deadlocks through a single virtual channel, which implies simple router design.

Similar to the technique used in [96], where credits are piggy backed with the data transfer, CTC allows opening in parallel other virtual connections for packet transfer, thus hiding connection set up latency during producer-consumer handshake. Thus, CTC is particulary advantageous for multi-producer communication, since it saves buffer space at the consumer side through buffer sharing among multiple data streams. In simple cases with a direct (virtual) producer-consumer connection, CTC can be configured to act efficiently without wasting for any connection set up latency.

As described in Section 4.1, language primitives for message passing models, such as send and receive, can be easily realized through a store communication primitive. However, due to tradeoffs between implementation efficiency and performance (e.g. network latency), we can argue that send and receive operations implemented using the push communication primitive are an attractive alternative to shared memory-based implementation.

4.7   Spidergon STNoC Building Blocks

The Spidergon STNoC IPU technology is a set of of communication primitives and low-level platform services on top of an on-chip communication network implemented using three different types of hardware building blocks appropriately interconnected according to a flexible and customizable pseudo-regular network topology.

As highlighted in Section 4.3, Spidergon STNoC IPU technology exposes programming models and middleware through its software framework, which forms the fourth (software) component. Next, in this Chapter, only the first three hardware blocks are described. Although the sofware framework will be examined in more detail in a future publication, a brief presentation is provided in Chapter 5.

Spidergon STNoC technology allows creating sophisticated Multicore SoC communication platforms by configuring and assembling three simple blocks in a LEGO block-like approach. Figure 4.19 shows these three Spidergon STNoC building blocks.

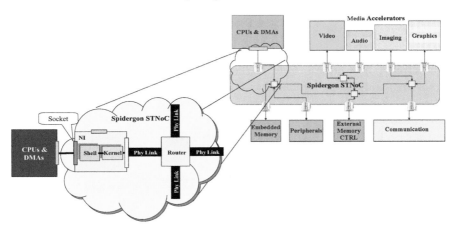

FIGURE 4.19: Spidergon STNoC: building blocks

The Spidergon STNoC network interface (NI) is the access point to the NoC, converting transactions generated by connected IP or subsystems into STNoC packets that are transported through the network. The NI hides network-dependent aspects, decoupling the transport layer from the network. This property allows IP blocks to be re-used without further modification independent of how the NoC architecture subsequently evolves.

NI supports different IP socket interfaces, e.g. AMBA AXI, or STBus protocol. The NI is the point where low-level platform services are partially implemented and controlled through the software framework.

The Spidergon STNoC router is responsible for traffic flow and QoS support across the network. Thus, at a high-level of abstraction, routers transfer flits from input to the outport ports, according to routing and QoS header information. The Spidergon STNoC router is designed to support the full range of topologies and routing schemes described in this Chapter.

The Spidergon STNoC physical links are responsible for actual signal propagation across the network. Each router or NI has a simple physical interface that will allow to easily plug (or unplug) a link and decide on the type of physical communication at the very late stages of the design flow.

Next, we discuss architecture details and design driving factors for the router, NI and physical link.

4.7.1 The Router

This Section focuses on the Spidergon STNoC router architecture, discussing motivations, main features, and key design alternatives.

The Router component is responsible for implementing the network layer of the Spidergon STNoC protocol stack. Thus, it must ensure a reliable packet transfer through the on-chip network, according to a proper QoS policy. From a very high-level perspective, a router is based on a crossbar switch with a given number of input and output ports (called router degree or -arity), and proper control logic. The core functionality of the router consists of deciding the output port where to forward the flit (after arbitrating possible conflict situations) and managing link-level flow control.

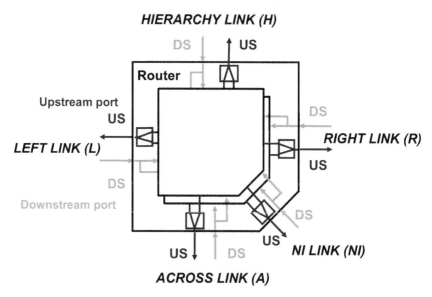

FIGURE 4.20: Spidergon STNoC router with upstream (US) and downstream (DS) interfaces

As shown in Figure 4.20, each Spidergon STNoC router implements wormhole routing and communicates through unidirectional links to a local NI, three neighbor routers in the right, left and across directions, and possibly another Spidergon STNoC network through a hierarchical connection. Thus, any hierarchical degree-3 chordal ring topology presented in Section 4.1 can be supported by implementing the routing schemes described in Section 4.4. As indicated in Figure 4.20, while input ports have a downstream interface, output ports have an upstream one.

The Spidergon STNoC router provides the capability to manage flows associated to two disjoint virtual networks that share physical link bandwidth.

Next, the main motivations and relevant features (driving architecture design and implementation) of the Spidergon STNoC router are given, in particular support for communication layering, configurability strategy, and performance goals.

4.7.1.1　Layers in Router Design

The traditional layered NoC design approach can be translated into a concrete strategy for router design. The router architecture must be end-to-end protocol-independent (and hence also independent of split transaction protocol details) and interface easily with different types of physical channel implementations. Spidergon STNoC applies layering to on-chip communication by identifying a specific block for each protocol stack level, while layer independence allows plugging different types of components to build better solutions for a specific system. From an industrial viewpoint, layering organizes on-chip communication as an open integration platform around which Multicore architectures can be composed. Moreover, layering focuses on independent optimizations and roadmap developments, such as CMOS technology evolution (in the link) or end-to-end protocol changes (in the NI).

To enable layering, the router architecture is fundamental, since all network control operations must be orthogonal to end-to-end protocol details and the data path logic must be able to easily interface different link solutions.

4.7.1.2　Configurability of the Router

The router block must be instantiated several times and properly connected to implement the defined on-chip network topology. Since overall interconnect flexibility depends on router configurability, the router must be a highly parametric block, providing modularity and allowing a fine grain tuning capability of the NoC micro-architecture at instance time.

Moreover, since cost is critical, several router parameters must be customizable to evaluate the best cost-performance tradeoff, for instance the buffer size, but not only. Unused virtual channels, physical links, and input and output ports could be removed, providing application-specific customization of the network architecture.

The Spidergon STNoC router is highly configurable, exposing its internal modular micro-architecture and allowing great flexibility in overall NoC design space exploration. A set of router parameters can be tuned at instantiation time to match functionality, performance and cost requirements. The router parameters are not necessarily changed in the same network instance, but they can let to different flavors of the same on-chip communication platform. Thus, when a complex network is being designed, a basic router can be configured and instantiated several times, but a different IPU instance may require a completely new basic router configuration.

Topology customization can be achieved through proper mirco-architecture configurability. As already explained, Spidergon STNoC has been designed as a flexible platform which is instantiated as a configurable whole, rather than as a collection of customizable router and NI components. To achieve this goal, each basic block is highly configurable. Only a small subset of the parameter space is exposed to the end-user, while other parameters are automatically inherited from the IPU instance.

The degree of flexibility to expose to the final SoC designer is an important architectural choice. In Spidergon STNoC, a decision has been made to allow low-level flexibility into the design of network building blocks by relying on the capability to partially hide complexity through a proper IPU tooling infrastructure. More specifically, in order to enable maximum NoC design flexibility in a user-friendly way, a top-level tool framework has been developed to support automated instantiation of all IPU building blocks from a high-level description. At router and NI instantiation process, the framework can be used to tune corresponding configuration parameters.

A first parametric option in the router is the flit size for each input/output link. These flit sizes must be checked with respect to other connected routers and NIs, to guarantee correct configuration. When a link is not used at all, since packets do not travel through it, the link can be removed, i.e. its port interface with all the relative internal hardware. Identification of unused links comes as a result of application mapping that is generally done for Multicore SoC. The size of all buffer resources in the router is configurable; this is especially important for low cost implementations when buffering can be selectively reduced depending on traffic type and performance requirements. Moreover, alternative queue architectures can be selected to increase global network performance, e.g. when lots of traffic is directed to a specific resource.

While Multicore architectures become more and more complex, with Spidergon STNoC IPU overall design complexity can be reduced. The Spidergon STNoC router has no externally accessible control registers. Its operation depends only on network configuration parameters defined upon instantiation and cannot be changed by software during runtime. This saves router area and programming complexity by avoiding switch initialization logic. Moreover, Spidergon STNoC services do not require any additional topological information since on-the-fly reconfiguration is performed efficiently at the boundary of the network, i.e. only in the network interface. This implies simpler on-chip network design, while still enabling efficient on-the-fly tuning capability.

4.7.1.3   Performance: Frequency, Latency, and Throughput

Performance is not only a matter of speed, but includes channel bandwidth, clock frequency per leakage, propagation latency, and network buffer organization. Properly setting these components can boost the accepted offered load, especially in the case of highly concurrent and irregular traffic. Latency depends on the router pipeline, but also on application traffic. On top of these parameters, QoS management for both network bandwidth and latency is fundamental for efficiently supporting complex applications with irregular communication patterns.

Clock speed is a driving factor in router design, although application-specific Multicore SoCs do not need GHz frequencies, due to physical implementation challenges, as well as power consumption, and cost reasons. An appropriate high-speed NoC design must utilize ultra-low power versions of CMOS technologies to provide better trade-off between performance and power consumption.

A low latency router is mandatory for ensuring fast communication. The depth of the router pipeline directly impacts the minimum latency, i.e. packet delay without any network conflict. In the presence of conflicts, latency depends on architecture, application traffic, efficient mapping onto NoC topology or virtual channel allocation. For example, access to a single external DDR memory can be a real system bottleneck. Despite these considerations, router latency is critical, especially for processor cores issuing cache refill operations through the network, requiring minimum response time. Also for this type of traffic, conflicts with other flows can change global duration of the transmission, thus availability of flow path independence, e.g. through virtual channel flow control, is strategic.

Spidergon STNoC router implementation proved to be an extremely cost-effective solution, allowing clock frequencies up to 1GHz in 65nm ST technology, with a 2-cycle latency, and a per link bandwidth of 8 GB/sec, assuming a 64-bit data path. Thus, the overall bandwidth for the router is 40GB/sec.

We next consider issues related to performance, such as buffer management and QoS.

4.7.1.4 Buffer Management

The buffering structure in the router and network interface design is a crucial concern in NoC design due to its tremendous impact on performance, area and overall on-chip network cost. Enough buffer resources are necessary for improving network throughput and reducing delay, especially for traffic bursts. It is also mandatory for supporting continuous packet transmission in a distributed multi-hop interconnect, hiding stop and go issues. Moreover, buffering ensures a simple back-end solution, since FIFOs can be seen as a retiming stage at the input or output interface of IP blocks.

Buffer management refers to various buffer organization schemes, packet arbitration and possibly virtual channel allocation algorithms. Notice that complex scheduling protocols with round-robin, static or dynamic priority assignments, furthest-first (overall or per algorithm phase), oldest-first, longest-first (for variable size packets), or combinations of these policies have worked well, especially in theory [208].

The exact buffer organization scheme, the number of buffers, and the buffer size are parameters decided during design space exploration. For example, a large buffer decreases network contention, while it increases packet delay, power consumption and switch area requirements. In general, the following buffer organizations are simple and can be used in NoC routers. Notice also that there are also combinations of schemes [161].

Input data buffering allocates one FIFO buffer for each virtual channel at each input port. Independent input queues provide simple and efficient flow control management, since they can be tuned based on link characteristics, e.g. latency and throughput. The packet to be transmitted through an output port can be selected from all input buffers in round-robin fashion. Any packet directed to a full buffer is stalled. Input buffering leads to the well-known head-of-line blocking problem, limiting throughput to ~58.6% for round-robin scheduling and 63% for random scheduling [98] [136] [149]. However, notice that simple parallel iterative matching approaches use distributed random scheduling to decompose the mapping from many-to-many to one-to-many and eventually one-to-one mappings, achieving performance close to an output buffered switch, with just 3 or 4 iterations [13] [245].

The split input data buffered switch (crosspoint or virtual output) implements FIFO buffers at every input port, one for each input-output port combination [98] [99] [100]. The switch can pre-route a packet to determine if it is directed to a full buffer at the next router. If all packets corresponding to a given output port are directed to a full buffer, traffic is temporarily stalled. Otherwise, the packet to be transmitted is selected using round-robin. The main problem with this switch is utilization of the buffer queues, when network traffic is non uniform.

The output data buffered switch implements FIFO buffering at the output ports [100]. All packets arriving simultaneously must be wired to the appropriate output buffer(s) before new packets arrive. The output buffering switch also uses pre-routing to determine whether packets are directed to a full buffer. All such packets are temporarily stalled. Thus, by using a FIFO buffer for each virtual channel at each output port the output buffering switch overcomes head-of-line blocking effects.

All switches above use statically allocated buffers, thus each buffer serves a single input stream. For better buffer utilization, more complicated switches, such as the central data buffered switch (or dynamically allocated fully connected), provide a dynamically shared data buffer for all virtual channels [100]. Central buffering design requires complex hardware data structures (pointers), especially for more than two input ports, but achieves optimal performance, i.e. similar throughput and reduced delay compared to output buffering. However, the dynamically allocated fully connected design fails to sustain acceptable bandwidth in the presence of non-uniform traffic spots, i.e. link contention for special communication patterns, since this type of switch places no limit on the number of slots a single input stream may take.

The central buffer flow control is per virtual channel, where virtual channel header information identifies uniquely the packet's producer and consumer. Round-robin is implemented at each output port, for selecting the packet to be transmitted. Simultaneous data buffer writes/reads upon virtual channel arrival/departure are possible through multiport or pipelined memory [179].

The dynamically allocated multi-queue buffering strategy allocates a central buffer and allows independent access to packets destined for each output port, while applying free space management to any incoming packet [350]. Due to complex control mechanisms, this scheme has higher area requirements than distributed dedicated buffering, but achieves higher performance due to better utilization of the available buffer space, especially for non-uniform traffic.

For very high performance protocols with hard real-time guarantees, priority-based crossbar architectures can be attractive. By providing virtual channels and separate buffers (or access schemes) for different priority packets, it is possible to allow a network to select high-priority packets to bypass others, and thus be delivered first. Since these schemes require additional buffers for each priority level, they are appropriate for high performance applications, e.g. in telecommunications that require quality of service, i.e. guaranteed throughput or latency.

The Spidergon STNoC router uses input and output FIFOs with the primary target to have independent control over efficient link transmission and global network throughput, while at the same time minimizing buffer requirements. Since buffer capacity required for flow control is independent of that for absorbing suspended traffic at each router, different buffer management techniques are provided. An output FIFO can be allocated for each output port, reducing head-of-line blocking. These queues are shared among input flows to avoid costly time/space speedup factors. Multiple queues for each incoming flow can be eventually instantiated in case of high traffic to a local node to avoid many conflicts for packets that must be ejected from the network. Moreover, output FIFOs may be completely removed if the given application traffic does not utilize them. FIFOs have bypass capability to reduce the pipeline, while an advanced output scheme supports virtual channels and at the same time offers a retiming capability.

Nowadays, using standard bus technologies to meet all backend requirements per each product causes several iterations during layout, extraction, and simulation within a short period of time. Spidergon STNoC router has been designed to support a new methodology during the backend phase. Each router is self-contained, i.e. all input and output ports are retimed, hence router logic delay is independent of link wire delay, allowing orthogonal optimization of front-end versus back-end. Thus, Spidergon STNoC router blocks may be reused as hard macros, saving months during the backend design phase. Futhermore, using fully characterized blocks greatly reduces the risk of errors.

4.7.1.5 Quality of Service

Quality of Service indicates the ways to manage bandwidth and latency to ensure a minimal requirement for each traffic flow. Quality of Service support is a global strategic service of the IPU which spans over the different network components, but must be properly solved at the router, since contention occurs at this block. Arbitration is different from allocation; requests asking for a single shared resource are arbitrated, while those asking for several alternative shared resources are allocated. The simple routing schemes and trivial virtual channel management in Spidergon STNoC router require concentrating on the arbitration problem.

Arbitration is a critical part of the router, since it determines the level of QoS support of the network and impacts router (and eventually nerwork) performance in terms of critical path delay. Two factors affect performance, the number of input request ports of the arbiter and complexity of the arbitration scheme. Another important issue is the capability for the router to be flexible enough to allow a certain degree of configurability of the global network arbitration policy.

The Spidergon STNoC router supports virtual channels on top of the same physical wire structure to maximize wire and gate efficiency. Virtualization can be used to design platforms where independent traffic classes share the same wires, while assigning proper priority levels.

As far as bandwidth is concerned, Spidergon STNoC supports the Fair Bandwidth Allocator (FBA) QoS mechanism. It is an end-to-end service that guarantees fair and programmable weighted bandwidth allocation on top of a distributed network, just by tuning the injection points. Similar to routing, FBA is a two-step QoS scheme that provides all necessary information in the network header, limiting router behavior to simple operations. Configuration is provided through NI registers that are programmable by the software framework (see Figure 4.5) without impacting network routers. Thus, for example, routing can change without any effort by re-computing the path followed by a flow and corresponding QoS parameters along this new path. Moreover, FBA is cost efficient, i.e. the router implements a simple arbiter, without complex and slow logic. Finally, the adopted QoS scheme follows a unique structured approach, whose flexibility is achieved by simple software control, without changing the hardware implementation; the same scheme can degenerate into simple schemes, such as round-robin, LRU, or priority-based.

4.7.1.6 Architecture of the Spidergon STNoC Router

The Spidergon STNoC router micro-architecture consists of data path and control logic distributed in input, switch and output stages. Figure 4.21 shows the architectural organization of the router and intends to clarify modular organization of the block. Input and output stages feature optimized pipelines that increase throughput, while flits and control signals may be latched through flip-flops (for retiming) or FIFOs (for buffering and retiming). This approach avoids introducing an extra delay in the internal critical path related to link wire delay. This orthogonality between link and router delay is fundamental to enable simple placement and routing during physical design.

Notice that although not directly highlighted in Figure 4.21, there is a clean interface with the physical link. The link is a stand-alone component in the Spidergon STNoC design. Different links can be plugged at each router port, thanks to a simple and universal physical hardware interface (few signals) that reflects the NoC protocol stack interface between layers 1 and 2.

The input stage manages incoming flits according to credit-based hop-by-

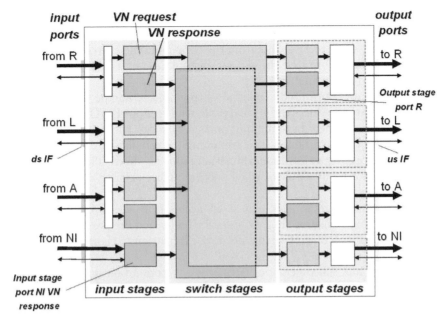

FIGURE 4.21:   Spidergon STNoC Router architecture

hop flow control. Since Spidergon STNoC does not allow flit drops or retransmissions, if a flit is transmitted, then at the destination there is always enough space to receive it. To guarantee maximum throughput, the input buffer is dimensioned according to credit round-trip delay, defined as the minimum time interval between two consecutive credits for the same buffer location. This delay depends on credit and data pipelines within the router and link; different types of Spidergon STNoC links result in different round-trip delays.

Then, the routing path of the incoming packet must be decoded by proper routing computation logic which examines the packet header already available in the first flit. When the corresponding output port is identified, the input stage must request access to an arbiter, since only one output virtual network can be accessed by a packet at a time. The routing block consists of fast combinatorial logic that determines the output port where a packet must be forwarded after decoding the destination and direction fields available in the first flit of the network packet header (to avoid wasted cycles). The logic provides flexibility to support different Spidergon routing algorithms by modifying appropriately the packet header field, i.e. programming the NI, without reconfiguring the router.

Switch stages connect input to output ports. An independent switch stage is provided for each virtual network. If the output link scheduler grants the transfer, then flits may move from input to output stage according to their routing computation. Notice, however, that reduction of router cost and complexity is possible if virtual channels are not instantiated in certain input/output ports, removing internal paths.

Each output stage receives flits from the corresponding switch stage. Then, the output stage arbitrates the output virtual channels and schedules flit transmission to the output link.

High frequency, optimized pipelined output stages handle outgoing flits. Parallel arbitration reduces the critical path, while output FIFO bypass is used to limit pipeline latency. Several input ports may access the same output port, and once a first flit is granted, the remaining packet flits can be transferred without any conflict, unless link arbitration stops the flow. This arises when credits are no longer available or when another higher priority virtual network wins link ownership.

The output stage must also request flit access from the link scheduler. A special mechanism ensures that requests for link scheduling are generated at least in parallel with output virtual network arbitration to shorten the critical path.

The output link scheduler prioritizes accesses from an output stage to the output link, using hop-by-hop credit-based flow control. Flits are sent to the output link only when credits are available in the input buffer of the adjacent router or NI.

More specifically, when the output stage has a single virtual channel, output link scheduling is based entirely on credit-based flow control, while for two (or more) virtual networks the output link scheduling is per-flit. Using virtual channels is a way to increase link utilization, while performance of flit interleaving on the link depends on credit availability and dynamically changing priority and is also critical due to reordering problems. A packet usually flows without any interrupt, reducing network latency.

As discussed for the input stage, if the target application does not use a particular output port, or if it requires limited performance, then corresponding hardware (e.g. the output virtual channel FIFOs) may not be instantiated, thus reducing cost.

4.7.2 Network Interface

In Section 4.1, we have introduced the concept of communication primitives and low-level platform services. These could be implemented in hardware or as a combination of hardware and software. The Spidergon STNoC Network Interface (NI) provides a hardware access point to external IP or processor cores and the necessary hardware to implement a set of communication primitives and low-level platform services. The two primary goals of the NI are to

- enable the connection to cores, converting high-level transactions generated by attached components into Spidergon STNoC packets, and

- partially implement low-level platform services.

The NI hides network-dependent aspects from the transport layer, allowing efficient support of evolving end-to-end protocols, while keeping an independent development of the interconnect service roadmap.

The Spidergon STNoC NI is the only hardware point of the IPU where implemented communication services are controlled through a set of accessible registers.

Depending on the programming model (e.g. shared memory, message passing, or data flow) and on the end-to-end protocol (e.g. OCP, AXI, STBus, custom memory interface, or FIFO-based streaming interface), the NI decodes and creates specific STNoC packets. Apart from protocol conversion the NI is in charge to adapt the frequency and the data bus size of the connected component to the rest of the communication platform.

This Section focuses on general architectural concepts adopted by the Spidergon STNoC NIs, considering the shared memory NI as a reference model.

4.7.2.1 Network Interface and Communication Primitives

In NoC domain, the NI provides a generic access point to computing resources. It forms a bridge between NoC routers and connected components, including memory banks and general purpose or specialized processing elements. Aggregation is one of the features of the NI, thus several components may be connected to a single NI.

In a shared memory context, the NI can be classified as master, i.e. requesting the memory transaction, or slave, i.e. performing the effective memory access and providing the response to the master component. The communication semantic is usually based on simple load and store operations implemented on top of request and reply messages.

Complex protocols dealing with cache coherence can be supported by introducing further communication primitives, e.g. with new packet types. Memory-based NIs support a minimal set of shared memory directives and provide hardware buffering to extract/inject network messages. Moreover, shared memory model at the network interface helps preserve the sequential paradigm of programming by imposing sequential semantics on writable shared values and simplifies emulation of message passing, e.g. by implementing transmit and receive queues as cacheable memory buffers.

More specifically, NI functions may include removing incoming network packets from the NI input ports, mapping (with possible frequency and size conversion) custom memory access requests to shared memory operations, initiating and executing shared memory transaction requests, creating and injecting replies directed eventually to the local computing resoucre, and implementing directory-based cache coherency protocols.

In the on-chip domain, NI provides interoperability through simple point-to-point communication channels, as well as existing standard on-chip protocols, such as AMBA AXI, OCP, or STBus, to seamlessly connect existing components supporting these protocols to the on-chip communication network. The capability to support different standards and enable heterogeneous communication is a mandatory requirement, strongly pushed by current and future Multicore platforms which are composed by components implemented by different organizations or companies.

Low-level direct network interfaces allow the processor to handle packets directly out of the network queues or with minimal buffering, thus further improving throughput and reducing parallel application latency through additional network hardware. Several machines have provided direct network interfaces, including CM-5 [209], iWarp [47], Alewife [4], and Wisconsin's Coherent NI [243] exploiting standard cache-coherence protocols when the NI is far from the processor.

Past research indicates that direct interfaces that can be accessed at cache speeds can offer better performance [141] [243]. Direct message passing in Multicore SoC is based on efficient mapping of the data flow, e.g. streaming applications, onto a homogeneous distributed architecture. The NI must support the direct transfer of data in a producer/consumer context, dealing in particular with potential deadlock issues related to several message dependencies.

The Spidergon STNoC communication semantic is quite generic to enable many types of protocols, since packets are transferred in point-to-point manner from source to destination. On top of this network level communication, end-to-end protocols can be easily mapped. The Spidergon STNoC technology provides network interfaces that support traditional shared memory communication with end-to-end protocols, such as the ST proprietary STBus or the AMBA AXI. Moreover, a NI with dedicated streaming interfaces and the capability to support direct message passing can be used to build an IPU with advanced programming models.

Intermediate solutions (direct data transfer through a shared memory semantic) can be designed using particular features of the Spidergon STNoC NI, e.g. through implementing distributed DMA support at each local network interface.

Direct memory access (DMA) is normally performed using a programmable engine moving a certain amount of data bytes, e.g. x bytes, from a starting address of a source memory to a destination address of a target memory. In practice, a DMA transfer, programmed by a central host, starts to load data from the source slave and then store them to the destination memory in a step-by-step process. Hence, two interconnect data transactions are necessary to complete the DMA.

A special Spidergon STNoC NI provides more efficient DMA transfers by allowing each memory to transfer data directly to another memory, exactly as a local (distributed) DMA, but requiring a single data transfer through the interconnect.

A central controller plugged to the Spidergon STNoC through a standard master NI can be the initiator of DMA communication by issuing a transfer directly through the NoC using a special opcode. The initiator of the distributed DMA request must provide the starting addresses of the source memory, the destination address of the target memory, and the transfer size. Without loss of generality, we can assume that the distributed DMA request is mapped into a particular write transaction.

The special opcode is eventually directed to the target slave, i.e. where DMA data must be transferred. This opcode describes an advanced operation that consists of issuing a NoC read operation from the source memory, collecting bytes, and copying them into the local memory starting from the destination address. Finally, the destination NI sends to the central controller a response through a special write response to acknowledge that the whole DMA has completed.

4.7.2.2 Network Interface and Low-Level Platform Services

The NI provides transport layer end-to-end management services that hide network dependent aspects from external components, such as connection handling, message segmentation and re-assembly.

The low-level platform services of the Spidergon STNoC IPU include QoS (to manage bandwidth and latency), power and security control, error monitoring, and reporting. As highlighted in Section 4.5.1.1, interoperability between IPs with different socket protocols is a crucial NoC feature, since the end user can integrate efficiently the IPU without using several bridges. More specifically, the NI is the point where to exploit protocol conversion, translating and adapting different socket signals into the flexible and user customizabe NoC protocol header.

In Spidergon STNoC, implemented services are supported by a sophisticated assembly of simple and efficiently optimized hardware functions which do not impact global network performance, enabling advanced application-based design. This means that network hardware is extensible, since unnecessary hardware is not instatiated.

Services must be reprogrammable, i.e. they are exposed to software for run-time on-the-fly reconfiguration via the low-level platform services (see Section 4.1). This application level flexibility is fundamental to accommodate multi-use case scenarios of current SoCs and mandatory in order to solve the complex multi-application management of future silcon platforms in the convergence market. Moreover, it offers a great opportunity to reduce bug and risk from time-to-market pressure, due to possible on-silcon, software-controlled performance tuning.

Exposing hardware platform flexibility to the software provides a great opportunity if it is designed with a pragmatic approach. Transforming hardwired parameters into accessible register sets is not enough for avoiding complex network re-programming. In Spidergon STNoC, the application associates a service to a given traffic flow, hence configuration software deals with per-flow service, rather than requiring complex network re-programmming. Moreover, Spidergon STNoC services are end-to-end properties, i.e. they are designed to be controlled at on-chip interconnect injection and extraction points, leading to trivial software management. The Spidergon STNoC NI is the only point of the IPU where services are controlled and exposed to application software, through a set of accessible registers coupled with a clean software API. The rest of the components are hardwired, defined at instantiation time, and their behavior is fixed, although globally the IPU is very flexible and run-time re-programmable. Programming only the boundary is a fundamental cornerstone in our vision of Spidergon STNoC as a scalable IPU acting as a unique, tunable and controllable system that hides topological details from the application.

The Spidergon STNoC peculiarity to implement network services at the boundary of the IPU provides benefits with respect to global NoC performance and micro-architecture efficiency. Supporting a network service, such as security management, has an implementation cost. It is much more efficient to associate this cost to the NI rather than to the router, since the NI does not deal with time-consuming router functions, such as routing, switching and arbitration.

4.7.2.3   Network Interface and Architecture Modularity

A fundamental issue in Spidergon STNoC NI design is modularity of the architecture and micro-architecture to maximize reuse and productivity. The main reason is related to protocol interoperability support: different processor cores must be connected to the network and the NI logic has to be reused across different core protocols. For example, the logic to manage packetization and de-packetization, or frequency and size conversion, or specific modules which implement the network services, must be independent of the type of connected core protocols.

Moreover, similar to the router, NI optimization focuses on configuring during instantiation the NI with a possibly reduced logic for traffic management, eliminating certain unused components. This is useful when a service is not required, or when the type of connected component emits only read (or write) traffic. NI buffers can also be set at instantiation time to a larger value to improve global network performance, with an extra cost in area and power consumption.

As previously discussed, the Network Interface is a re-programmable module at runtime through several service registers, but it is also very flexible in terms of its structure to allow the user to define a proper instance configuration.

There are two crucial separations in the Spidergon STNoC NI architecture that enable this modularity: a horizontal one which distinguishes the injection from the extraction path, and a vertical one which distinguishes between the network-dependent and the network-independent (connected component) part. These two parts are referred to as shell and kernel, as recently proposed in the design of Phillips Æthereal NI [280].

Separation between injection and extraction functions allows easy reuse of dual components in both master and slave NIs, since injection corresponds to packet composition and transmission, while ejection corresponds to packet reception and decoding.

Shell and kernel separation through relatively well-defined interfaces is really important for minimizing the effort of supporting different sockets, while keeping a fixed kernel structure and changing only the shell part. Moreover, this separation enables greater flexibility in the packet format that can be configured at instantiation time.

Since kernel deals with packet, while shell manages end-to-end protocol transactions, control and data signals are usually driven in parallel. Shell supports flow control to external bus protocols, while kernel handles NoC flow control at hop-by-hop and end-to-end level.

4.7.2.4 Architecture of the Network Interface

The architecture of a slave NI and a master NI are similar, except for a few details which do not change their global complexity and performance.

By definition, a master NI connects to a master component, thus initiating requests along its injection path and receiving responses from the master along its extraction path, while a slave NI connects to a slave component, thus receiving requests along its extraction path and initiating responses to the slave along its injection path.

As previously described and as shown in Figure 4.22, the Spidergon STNoC NI architecture decouples an end-to-end protocol-dependent shell from a NoC protocol-dependent kernel. The shell block manages end-to-end protocol interactions with external cores directly connected to the NI, while the kernel block is responsible for injection, extraction and temporary storage of STNoC packets. This decoupling idea stimulates reuse of independent parts when end-to-end protocol or NoC architectural requirements are modified.

Shell and kernel are connected by a simple and well-defined FIFO-like interface, which aims at ensuring a plug-and-play connection of different kinds of protocol specific shells.

Next, a detailed description of the shell and kernel modules is provided.

4.7.2.5 Network Interface Shell

The Spidergon STNoC NI shell connects the master (or slave) socket of the component to the NI kernel. The shell of a master NI is dual with respect to that of a slave NI. The first deals with requests in the injection and responses in the extraction path, while the latter one manages responses in the injection and requests in the extraction.

In general, the NI shell has to deal with socket component flow control, e.g. request/grant signals for STBus interface. The shell deals with the component data bus size and frequency, while potential adaptation in terms of size and clock speed is handled by the kernel part.

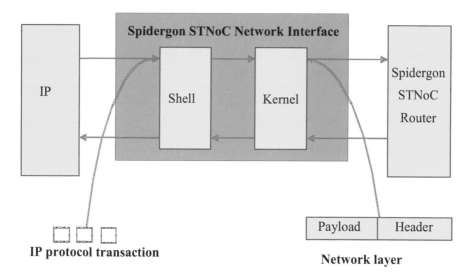

FIGURE 4.22: The STNoC NI top-level architecture

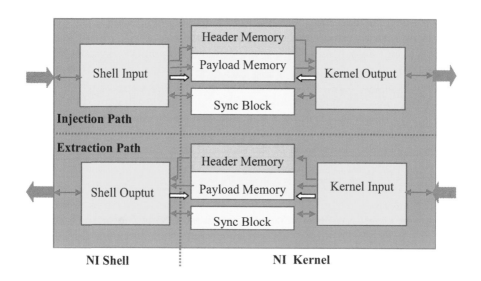

FIGURE 4.23: The STNoC NI shell and kernel block architecture

Along the NI injection path, the NI shell input block must manage storage of transaction data into kernel buffering, converting control information into an STNoC packet header and the potential data into an STNoC payload. As shown in Section 4.3.1, the Spidergon STNoC packet header is divided into network- and transport-level headers, to cleanly separate routing-specific information for transferring the packet from input to network output, and end-to-end information related to on-chip bus protocol semantics. Along the injection path, the shell is responsible for composing the packet header and encoding component socket information into Spidergon STNoC header fields. This encoding phase is controlled by software programmable parameters accessible through network interface registers.

In the extraction path the packets coming from the network are organized by the kernel buffering into header and payload, so the shell has to compose the end-to-end protocol transaction, decoding the header field and eventually collecting the data. The shell also uses a simple flag mechanism to tag transmitted flits in the injection or extraction directions to identify the transfer boundary without any expensive counter.

4.7.2.6 Network Interface Kernel

This Section focuses on NI kernel functionality and structure. As shown in Figure 4.23, two separate pairs of header and payload FIFOs in the Spidergon STNoC NI injection and extraction path temporarily store flits ready to be injected or extracted to/from the NoC.

As in the Spidergon STNoC router, NI buffers are organized and managed with flit granularity, but the user can decide to continuously inject packets to the network, avoiding wasted cycles. Using separate pairs of header and payload FIFOs decouples the shell from the kernel and provides significant advantages.

- It simplifies size and frequency management that is efficiently implemented though FIFO-based structures.

- Actual header size can differ from payload or flit size, so buffering can be optimized and reduced.

- In addition, for components that generate read- or write-only traffic, there is no need to have a payload FIFO in the master NI injection or (respectively) extraction path, thus reducing area complexity depending on the traffic type of the initiator component.

- Using a simple flag, the shell is able to simultaneously store packets while the kernel reads them without mixing flits of different packets in the master/slave NI injection path or equivalently in the master/slave NI extraction path.

Size conversion can occur on both the header and payload. For example, downsize conversion is necessary if the packet header is larger than the network flit size, since several flits must carry one packet header (the last header flit could contain stuffed bits). Similarly, downsize conversion occurs if the header flit received from the network is larger than the actual header size, and the kernel must discard unused (stuffed) bits. Depending on the relation between data bus size and network flit size, down- or up-size conversion may also happen at the payload FIFO in master and slave NIs, in both injection and extraction paths.

Since connected cores and network may run at different frequencies and share the internal NI buffers, the NI kernel must cope with frequency conversion. For safe buffer access, the FIFO control can be upgraded with synchronization logic able to ensure efficient asynchronous communication.

The kernel injection path is idle until a new header becomes ready to be sent, then the entire packet is transmitted flit per flit to the STNoC router, depending on network credit–based flow control. The flag mechanism drives the kernel to properly compose the packet with the right number of payload flits.

The dual task is performed by the kernel extraction path. Incoming packet flits are stored in either the dedicated header or payload buffers, and hop-by-hop flow control is managed depending on the availability of free locations in the NI FIFOs. This time the flag is not read, but properly set to drive the shell in downloading packets from the buffers.

Similar to the Spidergon STNoC router, the NI has a well-defined interface with the NoC link, thus any available Spidergon STNoC link can be plugged to implement NI-to-router communication.

4.7.3   Physical Link

The capability to integrate advanced physical link circuits in deep submicron technology is one of the key ingredients enabling efficient NoC design. With the scaling of CMOS technology, deep submicron effects complicate VLSI design, while wire delay increasingly affects system performance. Since deployment of clock distribution schemes is difficult, a lot of design time is devoted to resolving complex synchronization issues. Moreover, timing constraints often impose limitations to clock frequency, energy efficiency and architecture scalability.

Synchronization issues impose severe scalability limitations for deployment in NoC architectures. Several clock distribution strategies have been proposed that differ in assumptions concerning their clock frequency and phase [88].

- Traditional digital design uses synchronous systems which operate clocks that have the same frequency and phase shift.

- In pseudo synchronous (called pseudochronous) system design, the clocks have the same frequency and a known constant phase difference [253]; the constant phase difference is achieved using a digital programmable delay generator that allows control over phase differences between different clock regions.

- In mesochronous systems, clocks have the same frequency, but they are potentially out of phase with an arbitrary but constant phase difference [240] [232] [328] [382].

- Plesiochronous systems operate clocks with nearly the same frequency, and thus a phase difference that varies slowly.

- Periodic systems operate periodic clocks at arbitrary frequencies.

- Finally, asynchronous systems operate using signal events that may occur at arbitrary times and may use clocks with different frequencies and phases.

Nowadays, most NoC and relative point–to-point links are fully synchronous, i.e. all modules use the same clock, and flit transfer between any two connected routers occurs under the traditional synchronous assumption.

Frequency conversion exists only at the network interface between the NoC and core subsystems, thus enabling a globally asynchronous locally synchronous (GALS) approach. Although NoC is fully synchronous, the different connected cores can rely on multiple clock domains. Asynchronous inter-module communication among different synchronous modules is decoupled though NI frequency conversion; the global NoC is synchronous, thus this approach is fully supported by the current design flow.

While this simplifies NoC link design, distribution of a single clock on the entire NoC system (which aims at being physically spread on the chip) is a major challenge. To reduce clock skew is a complex design process that delays time closure and involves area and power costs due to the required clock buffers [108].

Asynchronous NoC links decouple clock domains, attempting to remove constraints on wire delays and reduce power consumption. Asynchronous delay insensitive schemes correctly recognize a code word independently of the delay of individual bits [60]. These codes can also be used for efficient NoC communication; however, efficient CMOS implementation (in terms of area and number of wires) is necessary [370] [27] [90] [25]. Since asynchronous design techniques are not supported by current design tools and standard design flow, the industry keeps on using the reliable and well-controlled synchronous paradigm.

Although recent research on fully asynchronous or de-synchronized NoC architectures (e.g. introducing random delays in a synchronous design) is not attractive for realization due to the above practical limitations, a pragmatic introduction of advanced link schemes is mandatory for eventual success of NoC as a physical distributed NoC platform.

As stated before, the main issue to overcome in the NoC context is efficient clock synchronization and distribution. Thus, a pragmatic solution addresses recovery of the skew, while considering the NoC operating at a unique clock frequency. Different clusters with different frequencies can be managed with standard conversion mechanisms, such as the one used in NI frequency adaptation.

The mesochronous approach could be employed to build reliable communication links providing high bandwidth channels for connecting NoC modules [232]. Mesochronous point-to-point link architecture enables communication between synchronous modules operating with arbitrarily skewed clock signals. The only requirement is that the transmitter and receiver clocks are derived from the same source. This means that although the clocks can be arbitrarily skewed, it is necessary that they hold the same phase relationship.

The mesochronous approach eliminates skew constraints in the clock tree synthesis, thus enabling frequency speedup, power consumption reduction and, above all, fast back-end turnaround time. Effective mesochronous solutions exhibit fewer limitations than asynchronous systems (such as high latency, high area, wire overhead and synchronization failures). The main techniques for state-of-the-art synchronization (using phase or metastability detectors) in mesochronous systems are discussed next. Notice that self-timed mesochronous interconnection schemes use components, such as voltage comparators, which cannot be implemented with standard cells [185] [131].

A cascade of flip-flops forms a simple so-called brute force synchronizer. Its operation principle is straightforward: the first flip-flop samples data and its output is sampled again by the second flip-flop in the next clock cycle. If metastability resolves itself within one clock cycle, the data are correctly latched. The probability of metastability propagation can be decreased exponentially by adding more flip-flops if a higher latency is tolerable. The brute force synchronizer can also be used as a component in complex synchronizers in mesochronous or asynchronous systems [185].

Periodic synchronizers identify transition points of periodic signals, like clocks that switch only at known points. Thus, it is possible to know in advance if it is safe to sample the data, or if it is necessary to wait. With mesochronous systems, phase shift remains constant, thus the periodic synchronizer can choose the right edge once and for all.

Delay line and two register synchronizers place a variable delay on each data line (or just clock) to allow synchronization phase to last an arbitrary amount of time, enough to guarantee a safe data sampling [83] [88] [90]. This delay is calculated to avoid switching in the metastability window of the receiving registers. The implementation uses a complex conflict detector that compares incoming data phase with clock phase to detect unsafe switching.

FIFO synchronization (or bisynchronous FIFO) is a common way to decouple components operating with multiple clocks in mesochronous (and asynchronous) systems. An m-stage buffer (flip-flops are sufficient only for 2-stages) and a complex initialization phase (for avoiding metastability) are used to latch incoming data according to the transmitter clock and to read data depending on the receiver clock [108]. Two pointers store the last written and last read location. In addition to avoid consistency issues, e.g. reading from an empty FIFO or writing to a full FIFO, complex logic based on expensive gray codes and brute force synchronizer is required to manage empty/full generation. In mesochronous systems, two stages are enough to absorb an arbitrary clock skew and wire delay, since there is a unique system frequency. The main drawback is that to avoid buffer size overestimation and area occupation and guarantee robustness against possible synchronization failures, the worst-case pattern must be extracted after extensive low-level design analysis.

Self-tested self-synchronization is a complex scheme for mesochronous systems based on failure detection [240]. This scheme uses two clocks with a phase shift of 180° (or more clocks with different phase shifts), that is the normal clock and its inverted copy. A complex failure detection process indicates which clock to use. The technique has a high area occupation, since it requires for each receiver a filter for detecting glitches, fast flip-flops operating at half the clock period and a circuit introducing artificial jitter in data lines during initial synchronization.

Finally, apart from global clock distribution and synchronization, wire delay is another critical issue that NoC links must address. Pipelined wires treat wire delay by interleaving relay stations (simple flip-flops driven by the global clock) along long wires [370] [115]. This method can resolve the problem of long global interconnections and achieve high throughput.

4.7.3.1 The Spidergon STNoC Physical Links

At the lowest level of the STNoC protocol stack, the STNoC physical link implements the physical layer of the NoC protocol. It is responsible for connecting routers to each other, and also routers to NIs.

There are several possible ways of implementing physical links, including combinations of synchronous/asynchronous and serial/parallel links. In fact, the choice of physical link technology involves tradeoffs between many issues, such as clock distribution, amount of on-chip wiring, and required chip area.

In this respect, decoupling provided by the layered Spidergon STNoC architecture is a major advantage, as changes to the physical layer can be subsequently made without affecting network and transaction layers. The clean link interface at each module port enables the capability to develop a link roadmap aware of physical and technological issues, and allows a simple mix at instantiation time of different link solutions within the same platform. The SoC architect composing the Spidergon STNoC IPU has to select the best link solution. Different link protocols can be plugged in the same NoC instance; for example, mesochronous can be used only in regions where clock distribution is critical in terms of skew.

Figure 4.24 provides a high-level view of the unidirectional Spidergon STNoC link, the well-defined interfaces and transmitter (upstream) and receiver (downstream) logic.

Adopted link-level flow control is a simplified version of credit-based. The

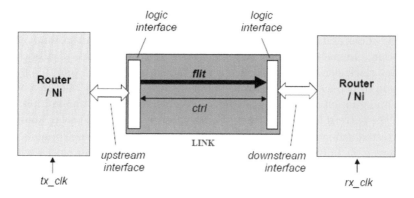

FIGURE 4.24:   Unidirectional link in Spidergon STNoC

basic link adopted in Spidergon STNoC is the synchronous parallel one, which has the advantage of being synchronous in respect to both the transmitter and the receiver, thus requiring a clock but no frequency adapter or synchronizer.

Based on the relay station concept, pipelined versions of synchronous link protocol are currently under development [59]. Asynchronous parallel links with a limited number of wires, i.e. using partial serialization, are under development for ad-hoc off-chip connection.

Links can be noisy channels, thus distributed or end-to-end error detection and correction strategies can be applied to the NoC at several levels of its layered architecture. A future Spidergon STNoC version will implement error detection and recovery techniques at packet and link level [290].

Next, we focus on the Spidergon STNoC mesochronous link called LIME (Link MEsochronous).

4.7.3.2   LIME: The Spidergon STNoC Mesochronous Link

LIME is an efficient point-to-point link that can be used in the Spidergon STNoC architecture, overcoming skew issues related to clock distribution. The LIME circuit is based on standard logic cells and is able to recover the skew and synchronize communication. Due to simple, unconstrained clock distribution, improved design productivity and reduced area and power consumption are expected. The LIME is an effective NoC communication channel, since it supports the Spidergon STNoC hop-by-hop flow control and virtual channel flow control with minimum buffer requirement and latency cost.

The shift from a synchronous link to LIME is transparent to the NoC modules due to fixed upstream and downstream interfaces that require adding few control signals on top of the physical wires, keeping essentially the same transmitter/receiver interface.

The global structure is extremely robust, reducing synchronization phase a single initialization step performed once to provide safe margins, while steady state behavior is shown to work by theoretical analysis.

LIME is a unidirectional link from an upstream module (transmitter interface) to a downstream module (receiver interface). This link consists of a data bus moving flits from upstream to downstream, with flow control supporting virtual channels, and control signals for low-level synchronization.

LIME is based on a smart combination of synchronization techniques: the bisynchronous FIFO and the ST proprietary SKew Insensitive Link (SKIL) mechanism [322]. This combination minimizes buffer requirements and latency, while ensuring that the best thoughput is equal to the maximum link bandwidth capacity.

The SKIL is an effective mesochronous synchronization scheme between modules operating with arbitrarily skewed clock signals; the phase relationship must remain constant in time among all copies of the clock. Some parameters can be tuned to guarantee robustness against jitter. SKIL can be implemented with standard cells using standard design flow.

The proposed solution does not require synchronization at each transaction, but only a small start-up time to adapt the phase relationship between the transmitter and receiver clocks. SKIL operation relies on a two-stage buffer to enable synchronization (fixed phase relationship) at startup time; in fact, the number of latches in the synchronizer circuit determines duration of the start-up time. Of course, during this short phase, transmit and receive operations are disabled.

Then, during steady-state operation, no further synchronization is required, guaranteeing robust reliability in subsequent communication by ensuring that the receiver reads only when data are stable. This is accomplished by SKIL's policy for writing and reading data into and from the buffer in a "ping pong" fashion. This is supported by a two-stage buffer structure which is written by the transmitter and read by the receiver. This structure has negligible overhead in terms of area and wire density.

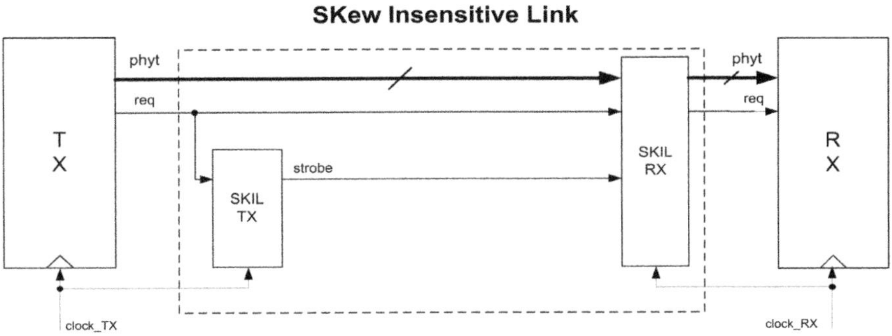

FIGURE 4.25: SKIL top-level architecture

Figure 4.25 shows a top-level view of SKIL. It is mainly composed of two units, SKIL_TX and SKIL_RX. The former provides the strobe signal needed at the receiver side for writing data in the buffer, while the latter includes the necessary buffering capability and manages the mechanism to recover synchronization at system start-up (through the strobe signal) and correctly read data from the buffer.

4.8   Tile-based Architecture

Growing system complexity and increased physical device and wire variability lead to conceiving new strategic approaches for developing computing architectures and design platforms. In Multicore system design, the fundamental issue does not concern the functionality mapped to each IP or processor core. Rather, it concerns the impact of communication requirements of different application functionalities. Communication-centric architectures based on novel IPU ideas, design methodology and tools deliver the next quantum leap forward in meeting manufacturing robustness and system-level predictability criteria at 45nm technologies and beyond.

In regard to design methodology, as discussed in Chapter 1.3.5.4, regularity is an important concept that increases productivity and helps resolve DFM problems, eventually providing higher yield. Regularity is usually based on simple modular patterns that can be applied at different abstraction levels, from system-level to RTL to implementation design flow of ASIC, FPGA, or programmable fabrics.

 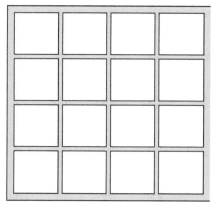

FIGURE 4.26:   Typical floor plan vs. layout of a regular tile-based SoC

Focusing on a modular topology, recently characterized tile-based architectures are based on regular on-chip communication networks, such as a honeycomb or 2-d mesh structure of switches and resources [89] [139] [195]. Each tile is regarded as an independent island, while symmetric network wiring is another intrinsic characteristic related to regularity in physical communication. Finally, interoperability among socket protocols relates to high-level regularity and provides the basis for component reuse. Figure 4.26 illustrates a SoC with an irregular layout on the left hand side and a completely regular floor plan on the right hand side.

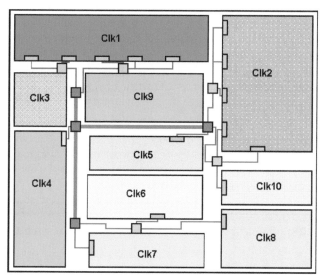

FIGURE 4.27: Typical clock synchronization in a regular tile-based SoC

As predicted in Chapter 1.1.1, future Multicore SoC will mix combinations of general-purpose CPU cores, specialized accelerator cores, and IPUs on a single silicon die. Although in this case, different tiles will not have the same ratio, and hence physical regularity cannot be guaranteed, tile-based architecture can still bring several benefits. For example, a possible solution to inter-tile communication can implement different links, e.g. synchronous, mesochronous, or asynchronous depending on physical constraints and communication requirements. In this case, as shown in Figure 4.27, tiles are completely decoupled in terms of operating frequency and voltage; notice that darker connections correspond to mesochronous links, while lighter connections correspond to synchronous links. Similarly, router links can be viewed as independent plug-and-play objects with well-defined interfaces. By adopting this modular methodology, time closure can be achieved faster by supporting well-known (semi) asynchronous protocols, such as GALS, in which "islands of synchronicity" are linked by asynchronous global connections, or LASGA, in which each tile dynamically adapts its own locally generated clock to local quantities, such as process quality, voltage, temperature, and transistor aging.

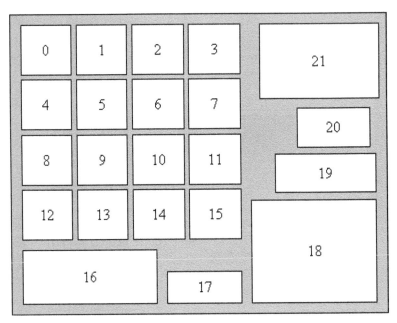

FIGURE 4.28:   Typical floor plan of a future multicore architecture

In certain application-specific domains, we can identify regularity at subsystem level, i.e. several cores are connected in a very regular fashion, e.g. IBM Cell or AMD Fusion which integrates a CPU with a GPU. Moreover, when the number of cores increases, e.g. in Intel's Tera-flop, it is natural to have a regular floor plan at system-level. However, due to coexisting commodity hardware components, such as DDR memory controller, USB, FLASH, and JTAG, which present their own constraints (e.g. placement next to the I/O pad ring), regularity is usually present only at subsystem level. Thus, a representative floor plan template for future Multicore architectures is shown in Figure 4.28; notice that modules 16 to 21 represent commodity hardware components, while modules 0 to 16 are processor cores.

Using the pseudo regularity of the Spidergon STNoC IPU, we can easily capture both layout regularity required by homogeneous Multicore processors (the same tile is replicated several times) and layout irregularity required by today's heterogeneous Multicore SoC. In both cases, a network interface is embedded within each tile and a router is placed between tiles. Thus, instead of routing design-specific global wires, inter-tile communication is achieved

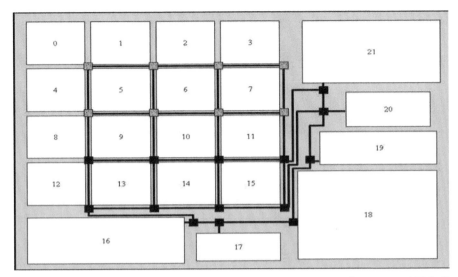

FIGURE 4.29: Eventual architecture floor plan on a 3-d Spidergon STNoC IPU with 22 nodes

through a communication backbone realized using these routers. For instance, Figure 4.29 illustrates a possible floor plan of Figure 4.28 implemented using a 3-d hierarchical Spidergon STNoC with 22 nodes (based on an 8- and a 14-node Spidergon) after removing unused components. Since the Spidergon STNoC IPU links are well-defined components, their type and electrical parameters can be controlled and optimized preferably later during backend design flow. Thus, the Spidergon STNoC IPU is a collective evolution of several technologies, with architectural innovation driving the capability to execute innovative applications in future Multicore platforms.

Chapter 5

SoC and NoC Design Methodology and Tools

5.1 SoC Design Methodology and Tools

A SoC is composed of many hardware and software components embedded on a single chip. These components can be high-level parameterized IP obtained from a library of reusable core and application software models. SoC design is a multi-disciplinary collaborative process addressing innovative, possibly domain-specific methodology, involving hardware/software partitioning, IP reuse, system-level design, digital and analog design, embedded systems, parallel and distributed systems, interconnection networks, application software, system integration, design for test, and verification.

Although system-level reuse and system evolution imply that 90% of hardware/software partitioning decisions can be made a priori, co-design issues are usually criticial, since they affect not only application execution time, but also power consumption and area requirements [191]. Unlike traditional manual hardware/software partitioning where hardware problems are resolved through software revisions, modern computer-aided co-design focuses on automated partitioning. Common hardware/software partitioning techniques minimize area while satisfying timing constraints, or vice versa. Algorithms are based on knowledge-based, dynamic programming, genetic algorithms, greedy heuristics and simulation annealing approaches [176]. In addition, notice that for Multicore there is also software-to-software partitioning, i.e. deciding how software tasks should be placed onto different (and possibly heterogeneous) cores. Hardware/software co-design reduces verification cost and time-to-market by considering simultaneously hardware and software architecture issues and performing early design assessment through executable, implementation-independent system models specified at multiple abstraction levels.

After hardware/software co-design, design flow can progress through co-specification of system-level models based on optimization, refinement and layering, co-synthesis (including architecture design and RTOS synthesis), co-simulation and co-verification of eventual hardware/software implementation.

In particular, SoC verification is one of the largest consumers of design time and effort. It includes formal verification, such as equivalence checking, and simulation-based techniques, such as debugging and tracing [112]. Current verification tools cover the whole spectrum of SoC design, including SW-based simulation with assertion checking, hardware emulation, e.g. using instruction set simulator (ISS), FPGA or core-based rapid prototyping, and hybrid methods. They also target post layout methods, such as coherent design for test plans using BIST or SCAN micro-architecture standards or standard interface wrappers, such as JTAG, DSM process analysis minimizing IR drop, signal integrity or power costs, OPC transformations for correct manufacturing [174], and post-fabrication testing.

Current design technology faces fundamental limitations inherent to the computational intractability of optimization in terms of the range of potential applications [52]. Existing design tools fail to exploit all possibilities offered by this technological leap, and shorter than ever time-to-market trends drive the need for innovative design methodology and tools. It is expected that a productivity leap can only be achieved by focusing on higher levels of abstraction, enabling optimization of the top part of the design where important algorithmic and architectural decisions are made, and massive reuse of pre-designed system and block components.

Thus, unlike existing RTL design and rapid system prototyping techniques, Electronic System Level (ESL) methodology is based on fast, power- and cost-efficient system-level (digital and possibly analog) models that are developed in a time-critical fashion using design and reuse principles [306]. Rapid evolution of ESL is enhanced by efficient analytical modeling and simulation, automated high-level synthesis, verification, and eventual testing. These models emulate expected application and system behavior through concurrent threads that compute, communicate and synchronize.

ESL focuses on the functionality and relationships of the primary system components, providing a global system-level view and separating system design from implementation. Low-level implementation issues greatly increase the number of parameters and constraints in the design space, thus extremely complicating optimal design selection and verification efforts. Similar to near-optimal combinatorial algorithms, e.g. traveling salesman or bin packing heuristics, ESL simulation models effectively prune away poor design choices by identifying bottlenecks and focus on closely examining feasible options through design space exploration, a key process to enhancing design quality. This is a key process in evaluating a vast number of design configurations in a time-critical fashion, generating new marketing ideas and shortening time-to-market [7] [97] [197] [198] [268] [275] [397].

It is obvious that several different NoC design tools are required for successful application-specific NoC design. These tools must deal with design methodology, as well as IP core selection and evaluation, application partitioning and mapping, job scheduling, NoC architectural simulation, design space exploration, including performance and power consumption estimation, concept validation, verification and possibly integrated hardware and software synthesis tools.

In this Section, we consider fundamental concepts in ESL modeling, system-level design methodology and architectural exploration adopted by the SystemC library and several system-level SoC design environments and tools. This standard design flow methodology aided by configurable IPs, appropriate application traffic characterization, and efficient debugging, profiling and visualization tools will foster NoC integration and parameterization.

More specifically, this Section is organized as follows. In Section 5.1.1, we introduce SoC modeling and discuss different abstraction levels. Then, in Section 5.1.2, we concentrate on transaction-level modeling (TLM) for efficient system-level design, and examine a recently proposed standard SystemC TLM API. In Section 5.1.3, we focus on user-driven top-down and bottom-up SoC refinement based on orthogonalization of concerns, i.e. separation of specification from architecture implementation, and module behavior from communication interface. In addition, for optimal refinement and independence of drivers from interconnect and resource models, inter-module communication layering is implemented. This methodology makes SoC modeling independent of the final architecture, enables extensive and systematic block- and system-level IP reuse, high-level application traffic modeling, transaction-level system modeling and hardware-software co-design by focusing at progressively lower levels of abstraction, and offers efficient design space exploration compared to traditional hardware description languages, such as VHDL and Verilog. In Section 5.1.4, we examine system-level performance modeling, introducing a SystemC-based performance framework based on objects that automatically extract instant and duration values from system components, and visualization components based on advanced monitoring. This framework has been implemented within the open-source On-Chip Communication Network (OCCN) framework [80] [82]. In Section 5.1.5, we focus on high-level design space exploration and design flow. Finally, in Section 5.1.6, we examine past and existing commercial and prototype system-level design tools and environments for Multicore design.

5.1.1 SoC Modeling and Abstraction Levels

A fundamental issue in system design is model creation. A model is a concrete representation of IP functionality. In contrast to component IP models, a virtual platform prototype refers to system modeling. Virtual platform enables integration, simulation, and validation of system functionality, reuse at various levels of abstraction, and design space exploration for various implementations and appropriate hardware/software partitioning. A virtual prototype consists of

- hardware modules, including peripheral IP block model (e.g. I/O, timers, Audio, Video code or DMA), processor emulator via ISS (e.g. ARM V4, PowerPC, ST20, or Stanford DLX), and communication network with internal or external memory, (e.g. bus, crossbar, or network-on-chip),

- system software, including hardware dependent software, models of real-time operating system (RTOS), device drivers, and middleware, and

- environment simulation, including application software, benchmarks, and stochastic models.

SoC modeling can achieve a higher degree of productivity through IEEE's SystemC standard, a popular open-source C++-based system-level modeling and simulation library [345]. SystemC consists of a collection of C++ classes describing mainly hardware concepts and a simulation kernel implementing the runtime semantics. It provides all basic concepts used by HDLs, such as fixed-point data types, modules, ports, signals, time, and abstract concepts, such as interfaces, communication channels, and events. A detailed system-level modeling framework can rely on a SystemC-based C++ Intellectual Property (IP) modeling library, a powerful simulation engine, a runtime and test environment, and refinement methodology. SystemC supports design abstraction at the RTL, behavioral and system level, allows development and exchange of system models and provides seamless tool integration from a variety of vendors. SystemC also collaborates with SpecC on synthesis. SpecC is a C++-based IP-centric co-design methodology and language enabling system-level executable specification, modeling, simulation and hardware-software co-design of embedded systems [116]. SpecC is currently supported by the SpecC Technology Open Consortium, which maintains a language reference manual and a reference compiler.

Several commercial tools for domain-independent SystemC-based design have been previously designed. Despite current system design efforts, there is not yet a complete and efficient SystemC-based development environment

- providing parameterized architecture libraries, synthesis, and compilation tools for fast user-defined creation and integration of precise and consistent SystemC models,

- enabling IP reuse at the system and block level which results in significant savings in project costs, time scale and design risk [111] [291], and

- supporting efficient design space exploration.

Detailed VHDL or Verilog models are inadequate for system-level description due to large system design and implementation complexity, unavailability of details during early design phases, and poor simulation performance. Thus, system-level modeling is performed at higher abstraction levels by using appropriate abstract data types (ADTs), enabling component description with object-oriented design patterns. These virtual SoC prototypes may hide, modify or omit system properties, thus allowing the designer to focus on IP functionality, rather than on detailed data flows or hardware details of the physical interface. Hence, abstraction levels in SoC modeling trade multiple levels of accuracy (measured in terms of model details) with simulation speed. An intuitive description for the most commonly used abstraction levels, from the most abstract to the most specific, is given hereinafter, i.e. from functional to transactional cycle- and bit-accurate to gate-level [137] [345].

- Functional models have no notion of resource sharing or time, i.e. functionality is executed instantaneously and the model may or may not be bit-accurate. This layer is suitable for system concept validation and functional partitioning between control and data. This includes definition of abstract data types, specification of hardware or software (possibly RTOS) computation, synchronization and communication mechanisms (e.g. using FIFOs), and algorithm integration.

- Transactional behavioral models (denoted simply as transactional) are functional models mapped to a discrete time domain to reflect timing constraints of the specification and delays of the architecture. Transactions are atomic operations with their duration stochastically determined, i.e. a number of clock cycles in a synchronous model. Although general transactions on buses cannot be modeled, transactional models are fit for modeling pipelining, RTOS introduction, basic communication protocols, test bench realization, and preliminary performance estimation.

- Transactional clock accurate models (denoted transactional CA) map transactions to a clock. Communication between modules is modeled using function calls called Interface Method Calls (IMC) [345]; this is accurate in terms of functionality and timing. Thus, synchronous protocols, wire delays, and device access times can be accurately modeled. Using discrete-event systems, this layer allows for simple, generic, and efficient cycle-accurate performance modeling of abstract processor core wrappers (called bus functional models), bus protocols, signal interfaces, peripheral IP blocks, ISS, and test benches. Time delays are usually back-annotated from RTL models, since Transactional CA models are generally not synthesizable. A more precise definition of transaction-level models is provided in Section 5.1.2.

- RTL models accurately reflect registers and combinatorial logic of the design at a detailed abstraction level so that synthesis tools can generate gate-level descriptions (or netlists). Register-transfer level systems are usually visualized as having two components: data and control. The data part consists of registers, operators, and data paths, while the control part provides the time sequence of signals that evoke activities in the data part. Data types are bit-accurate, interfaces are pin-accurate, and register transfer is time-accurate. Propagation delay is usually back annotated from gate models.

- Gate models are RTL models with additional information. They are described in terms of primitives, such as Boolean logic (i.e. registers and combinatorial logic) with timing data and layout configuration. For simulation reasons, gate models may be internally mapped to a continuous time domain, including currents, voltages, noise, clock rise and fall times. Storage and operators are broken down into logic implementing the digital functions corresponding to these operators, while timing for individual signal paths can be obtained.

Thus, according to the above, an embedded SRAM memory model may be defined as

- a functional model in a high-level programming language, such as Ada, C, C++ or Java,

- a transactional CA model, allowing validation of its integration with other components,

- implementation-independent RTL logic described in VHDL or Verilog,

- as a vendor gate library described using NAND, flip-flop schematics, or

- at the physical level, as a detailed and fully characterized mask layout, depicting rectangles on chip layers and geometrical arrangement of I/O and power locations.

In order to provide system performance measurements, e.g. throughput rates, packet loss, or latency statistics, system computation (behavior) and communication models must be annotated with an abstract notion of time. Analysis using parameter sweeps helps estimate the sensitivity of high-level design due to perturbations in the architecture, and thus examine the possibility of adding new features in derivative products [61] [60]. However, computation or communication delays are not always cycle-accurate and the designer must ensure correct behavior independent of delays by including all necessary synchronization points and/or interface logic. For example, for computational components mapped to a particular RTOS or large communication transactions mapped to a particular shared bus, it is difficult to accurately estimate thread delays, which depend on precise system configuration and load. Similarly, for deep sub-micron technology, wire delays that dominate protocol timings cannot be determined until layout time.

5.1.2   Transaction-Level Modeling for System-Level Design

As SoC and Multicore devices become increasingly complex, enhanced design flow demands a growing use of system-level simulation and analytical modeling, block- and system-level reuse and efficient tools for early design space exploration and concept validation for different architectural choices and customizations. One of the biggest design challenges concerns the development of a powerful, low cost and fast modeling and simulation environment for system-level design, and especially processor core, memory systems and network communications. System-level design flow is modular and progresses through algorithm description, optimization and verification, hardware/software partitioning, and system verification. Transaction Level Modeling (TLM) is a lightweight executable functional specification model for system-level design exploration and verification that captures intended behavior of integrated hardware and software architecture IP models at an abstraction level higher than traditional hardware description languages, without full-scale implementation details of the the physical interface, including data flows, FIFO size, or timing constraints [266] [56] [54]; these IPs can be obtained through a reusable library or developed by different design teams. Thus, TLM provides an early platform for embedded software development and architectural exploration.

TLM spans many modeling layers between the cycle- and bit-accurate micro-architecture model and untimed algorithmic models, trading simulation speed and efficiency in verification (at least 1000 times faster than RTL) with accuracy [349]. Furthermore, TLM provides a virtual prototype that reduces time-to-market through early system integration, verification, design space exploration and embedded software development with possible cross-compilation on the processor core, even before an RTL model becomes available. Moreover, when the TLM and elaborated transaction-level golden test bench models include back-annotated timing information from the equivalent RTL model, they act as a reusable reference model for RTL implementation and verification; in particular, reuse of the same test bench environment across different levels of abstraction is a fundamental issue for seamless equivalence testing that can be achieved by integrating special property specification languages for assertion-based verification, e.g. Sugar, e, Open-Vera. Thus, unlike FPGA prototyping and VHDL/Verilog netlists, SoC TLM platforms can be delivered shortly after the architecture is specified, promoting interoperability and enabling hardware/software (or customizable processor) co-design, integration and co-verification for real-time constraints to start very early in the design. Early analysis tools enable SoC designers to take advantage of the many possible system configurations available prior to the existence of a complete and detailed system description. During this phase, key tools help explore different design alternatives, evaluate insights or configuration parameters and identify potential bottlenecks prior to committing to an implementation by rapidly obtaining accurate metrics for SoC performance (throughput, latency and jitter), resource utilization and energy consumption. Notice that placement and routing is still needed to determine other system metrics, such as the number of wiring layers, clock speed, network congestion and chip area.

TLM modules, possibly shared among many design projects, exchange protocol-defined transactions through IMCs. A transaction is an atomic event specifying module interactions through a communication channel, such as data exchanges. It encapsulates a potentially complex data structure transmitted according to a protocol that may involve several low-level information transfers in both directions (data sent, acknowledgment, etc.).

TLM integration in the design flow, especially top-down and bottom-up design or synthesis, is not yet systematically developed and standardized [349]. Nevertheless, IP and EDA tool providers help TLM proliferate by adopting standardized or proprietary TLM APIs, e.g. SystemC, OCP-IP or Cadence formats, for fast and efficient simulation and eventual synthesis of protocol-specific IPs, such as AMBA or OCP bus read/write transactions. Similarly, a wide variety of powerful SoC design tools are currently being developed for extended debugging, efficient simulation and visualization, commercial and non-commercial TLM-to-RTL and RTL-to-TLM translation, hardware/software co-design, high-level synthesis and verification.

5.1.2.1 System-Level Design using TLM API in SystemC

Open TLM standards reduce vendor dependency and increase reuse since a non-proprietary technology can be easily supported by all IP providers. TLM is supported by SystemC 2.0 (a C++ library that has recently become IEEE standard 1666) and SystemVerilog. While SystemC addresses modeling of complete SoC composed of hardware and software components and provides superior features at higher levels of abstraction, SystemVerilog focuses more on hardware design and verification. However, none of the languages provides dynamic reconfiguration or reuse principles.

SystemC supports development of TLM models very well by providing an adaptable and efficient communication infrastructure based on modules, events, interfaces, channels, ports and signals. A module is a structural element which contains processes, signals, events, interfaces, channels, ports, clocks and other modules. An interface defines a set of methods, but does not implement these methods, i.e. it is a pure functional object. By distinguishing the declaration of an interface from the implementation of its methods, SystemC promotes a coding style in which communication is separated from behavior, a key feature to promote refinement from one level of abstraction to another. A channel implements one or more interfaces. With channels, there is a distinction between so-called primitive channels and hierarchical channels. Primitive channels do not exhibit any visible structure, do not contain processes, and cannot directly access other primitive channels. Hierarchical channels, on the other hand, are modules, which means they can contain other modules and processes and directly access other channels [345]. A port is defined in terms of an interface type, which means that the port can be used only with channels implementing that interface type. A port acts as an agent that enables a module process to obtain communication functionality by accessing a channel interface. Future SystemC language features will include dynamic thread creation, forks, joins, interrupts, aborts and support for timing constraints and real-time modeling, including scheduler modeling.

Since communication is modeled using SystemC channels, transaction requests are implemented through calling interface functions of these channels. Transaction-level models often implement atomic communication actions using simple and uniform tokens. At the transaction level, emphasis is more on functionality of the data transfers rather than actual implementation [56] [313] [312]. Thus, for example, AMBA APB and AHB bus transaction level models run at high speed, while complex RTL bus models with accurate hardware signal descriptions are much more CPU intensive [82]. The current TLM OSCI standard defines several interfaces for transaction-level communication, the most important ones being blocking and non-blocking ones, e.g. put and get. These standard SystemC interfaces provide the groundwork for the definition and implementation of clock-accurate, loosely-timed and pure functional TLMs of systems.

Moreover, OSCI's TLM Working Group has recently released new draft specifications of its TLM 2.0 standard for public review. This draft also defines the content of transactions, i.e. a generic payload describes data structures for address and data transmissions. This concept simplifies exchange of transaction-level IP by avoiding the design of wrapper components, and makes transaction-level models interoperable across different companies, design groups, and vendor EDA tools.

5.1.3 General System-Level Design Methodology

Formal refinement relates to properties which must hold between the abstract and the refined model [93]. A refined model is systematically translated from an original model by maintaining correct system functionality, but differentiating from it through new communication, computation, or synchronization properties specified in different specification languages, using temporal logic systems. Properties are model attributes and involve structural, e.g. and memory size, or functional elements, e.g. data consistency and protocol behavior. Micro-properties describe basic relations among attributes, while macro-properties are collections of micro-properties. Thus, satisfaction of a macro-property ensures all corresponding micro-properties are also satisfied, but not vice versa. Properties are also categorized into three distinct classes.

- Structural properties refer to both system performance and correctness, e.g. FIFO and memory size, ALU operand size and instruction size. These properties concern implementation and appear at lower abstraction levels.

- Functional properties mainly refer to correctness, e.g. mutex and data consistency.

- Performance properties mainly refer to performance characteristics, e.g. throughput, latency, jitter and power estimation considering NoC topology, data flow, and communication protocol design.

Optimal system modeling methodology is a combination of stepwise protocol refinement, hierarchical modeling, orthogonalization of concerns, and communication layering techniques. A crucial part in the stepwise transformation of a high-level behavioral model into actual implementation is refinement [233]. Stepwise protocol refinement is achieved through a top-down and bottom-up design automation approach that attempts to improve design productivity.

- In bottom-up refinement, IP-reuse oriented integration with optimal evaluation, composition and deployment of pre-fabricated or pre-designed IP block and system components drives the process. Component-based design comes under this design approach. It starts with a set of components, possibly selected from IP libraries, such as IBM's BlueLogic [159] and Synopsys' DesignWare Star IP [343], and provides a set of primitives to build fully-programmable application-specific SoC architectures and communication APIs. Thus, component-based design promotes IP reuse and modularity through support of multiple interoperable levels of abstraction, protocol refinement, compatibility testing and automatic verification.

- In top-down stepwise refinement, emphasis is placed on specifying unambiguous semantics, capturing desired system requirements, optimal partitioning of system behavior into simpler behaviors, and gradually refining the abstraction level from system specification down to a concrete low-level hardware and software architecture by adding details and constraints in a narrower context, while preserving desired properties. Top-down refinement allows the designer to explore modeling at different level of abstractions, thus trading between model accuracy with simulation speed and increasing design productivity dramatically. The above continuous rearrangement of existing IP in ever-new composites is a key process to new marketing ideas. It also allows for extensive and systematic reuse of design knowledge and application of formal correctness techniques. The concrete architecture obtained through formal refinement must satisfy relative correctness, i.e. it must logically imply the abstract architecture, and faithfulness, i.e. no new properties could be derived. High-level (behavioral) synthesis (HLS) from system-level models is a popular and promising top-down design. HLS automates implementation of functions and algorithms, but it is currently difficult to apply globally on a complex Multicore. Platform-based design (PBD) is a top-down integration, verification and interoperability-oriented design approach that provides a continuous and complete design flow from hardware/software partitioning to RTL generation. With PBD, complex products are built using kernel components, such as processor cores, RTOS, NoC interconnects, memory and standard interfaces, as well as qualified embedded software IP (called virtual components) configured appropriately based on algorithmic, architectural and technology choices

[181] [273] [299] [298]. PBD aims to reduce development risks, costs and time-to-market by emphasizing systematic IP reuse of internal or externally sourced, programmable qualified IP blocks of compatible domain-specific IP, e.g. based on SPIRIT [330], VSIA [373] or OpenMore quality standards. Along with fast and easy integration of IPs, PBD facilitates rapid creation and verification of designs by automating complex, tedious, and error-prone design creation, IP integration, and verification steps.

While formal refinement is hard to achieve, since there are no general proof techniques, relative refinement is based on refinement patterns consisting of a pair of architectural schemas that are relatively correct with respect to a given mapping. By applying refinement patterns, e.g. a general optimized state transformation [239], we can systematically and incrementally transform an abstract architecture to an equivalent lower-level form. Notice that formal architectural transformation is related to providing an augmented calculus with annotations for properties, such as correctness, reliability, and performance [226]. If architectural components eventually become explicit formal semantic entities, then architectural compatibility can be checked in a similar way to type checking in a high-level programming language [8].

Hierarchy is a fundamental relation for modeling conceptual and physical processes in complex systems of organized complexity, based either on a top-down analytical approach, i.e. the divide and conquer paradigm, or a bottom-up synthesis approach, i.e. the design-reuse paradigm. Hierarchical modeling based on simpler, subordinate models controlled by high-level system models enables a better encapsulation of the design's unique properties, resulting in efficient system design with simple, clean, self-contained, and efficient modular interfaces. For instance, hierarchical modeling of control, memory, network and test bench enables the system model to become transparent to changes in sub-block behavior and communication components. A systematic framework supporting hierarchical modeling may be based on the SystemC module object, and the inheritance and composition features of the C++ language [65] [371]. Thus, SystemC modules consist of SystemC sub-modules in nested structures. Besides, SystemC facilitates the description of hierarchical systems by supporting module class hierarchy very efficiently, since simulation of large, complex hierarchical models imposes **no** impact on performance over the corresponding non-hierarchical models.

Object-oriented system design is based on virtual micro- and/or macro-architectural components (called IP), interacting with each other in a specific system environment. Within each IP, we distinguish between behavior and communication components [181].

- A behavior (computing) component describes module functionality. It usually has an associated identity, state and an algorithmic process consuming or producing communication cells, synchronizing or processing data objects. Access to a behavior component is provided via a communication interface and explicit communication protocols.

- Inter-module communication (protocol interface) specifies what the component expects from the environment for achieving the desired system functionality. This usually refers to I/O ports transferring messages between one or more concurrent behavior components. Each communication port is connected using an interface to a communication channel object. The interface may support different communication protocols and is the only way to interact with the behavior. It must also support an optimal design methodology based on top-down and bottom-up refinement. Thus, we fully de-couple behavior from inter-module communication.

Behavior and communication interfaces can be expressed at different levels of abstraction. Static, time-independent behavior is specified with untimed functional models, while dynamic, time-dependent behavior is based on complex control, e.g. hierarchical Finite State Machines (FSMs) or Threads. Similarly, communication can be abstract or close to implementation, i.e. using generic interfaces, VCI [373], OCP [137], or proprietary interconnects [334] [335].

Stepwise transformation of a high-level behavioral model of an embedded system into actual implementation is naturally based on the concept of orthogonalization of concerns used in both research and commercial tools.

- Separation of system function specification from final architecture, i.e. what are the basic system functions for system-level modeling vs. how the system organizes hardware and software resources in order to implement these functions. This concept enables system partitioning and hardware/software co-design by focusing at progressively lower levels of abstraction. Notice that system function refers to models of computation and communication, such as dataflow, discrete event, concurrent finite state machines, message passing and shared memory systems, while architectural structure includes configuration of SoC component models, including processor, memory and hardware blocks.

- Separation of computation (called behavior) from communication enables plug-and-play system design using communication blocks that encapsulate and protect IP cores. This concept reduces ambiguity among designers, and enables efficient design space exploration.

Through successive functional decomposition and communication refinement, advanced system-level SoC modeling leads from a high-level model derived from initial specifications to

- a functionally correct protocol, guaranteeing consistency in respect to behavior, communication and system properties specification,

- an unambiguous protocol, whose original behavior is preserved in lower abstraction levels, and

- an optimized protocol, exploring performance vs. cost tradeoffs in the design space determined by all possible mappings from behaviors to architectural topology (called hardware/software co-design), while using a minimal amount of physical resources.

Layering simplifies individual component design by defining functional entities at various abstraction levels and implementing protocols to perform each entity's task. Thus, layering reduces ambiguity in writing communication drivers and enables plug-and-play system design by providing appropriate communication interfaces, encapsulating and protecting IP cores. Advanced inter-module communication refinement may be based on establishing distinct communication layers, thus greatly simplifying the design and maintenance of communication systems [51] [82]. Usually two layers are adopted:

- The communication layer provides a generic message API, abstracting away the fact that there may be a point-to-point channel, a bus, or an arbitrarily complex network-on-chip.

- The driver layer builds the necessary communication channel adaptation for describing high-level protocol functionality, including compatible protocol syntax (packet structure) and semantics (temporal ordering).

Using refinement and layering, behavior and communication interfaces are eventually mapped to hardware/software resources of a particular architecture. This categorization, known as system partitioning, is an essential element of co-design. Behaviors mapped to hardware logic are either synthesized or selected from existing IPs, e.g. a processor core. Behaviors mapped to software are assigned to a software process or device driver. Similarly, communication protocols are mapped to hardware, or software depending on available semantics, e.g. shared memory, message passing, Ada-like rendezvous, or queuing structure.

When the optimal mapping of behavior and communication components to architecture is reached, the designer may either manually decompose hardware components to the RTL level of abstraction or use available behavioral synthesis tools. Much of the hardware implementation can be reused from previous designs, including processor cores, bus interfaces and protocols, and large blocks, such as MPEG decoders. Similarly, on the software side, RTOS, device drivers and large blocks of code-like protocol stacks can also be reused.

5.1.4 System-Level Modeling and Performance Evaluation

In this Section, we consider the SystemC modeling library and simulation kernel; a proprietary runtime and test environment may be provided externally [293]. We examine high-level modeling based on user-defined C++ building blocks (macro functions) that simplify system-level modeling, enable concurrent design flow (including synthesis), and provide the means for efficient design space exploration and hardware/software partitioning.

High-level SystemC-based modeling involves hardware objects, as well as system and application software components. Since SoC design becomes software-intensive with improved performance, SystemC provides greater functionality and safety by relying on multiprocessing, dynamic multithreading with preemption, and scheduling. Hence, for models of complex safety critical systems, software components may include calls to a real-time operating system (RTOS) model [228] [121]. An RTOS model abstracts a real-time operating system by providing generic system calls for operating system management (RTOS kernel and multi-task scheduling initialization), task management (fork, join, create, activate, interrupt, sleep and abort), event handling (wait, signal, notify), and real-time modeling. SystemC 3.0 is expected to include user-defined scheduling library constructs and real-time operating system emulation characteristics on top of the core SystemC scheduler, providing RTOS features. During software synthesis, an RTOS model may be replaced with a commercial RTOS.

Focusing on hardware and especially on-chip communication components, we describe next system-level modeling objects (components) common in SystemC models. Most objects fall into two main categories: active and passive modeling objects. While active objects include at least one thread of execution initiating actions that control system activity during the course of simulation, passive objects do not have this ability and only perform standard actions required by active objects. Notice that communication objects are neither active nor passive. Each hardware block, is instantiated within a SystemC class, called `sc_module`. This class is a container of a user-selected collection of SystemC clock domains, and active, passive and communication modeling objects, such as

- SystemC and user-defined fixed data types, such as Bits and Bytes, providing low-level access and allowing for bit-accurate, platform-independent modeling,

- user-defined passive memory objects, such as Register, FIFO, LIFO, circular FIFO, Memory, Cache, as well as user-defined collections of externally-addressable hierarchical memory objects,

- SystemC and user-defined intra-module communication and synchronization objects, based on concurrent shared memory or message passing, such as Mutex, Semaphore or Event. This includes Timer and Watchdog Timer for generating periodic and non-periodic time-out events,

- active control flow objects, such as SystemC processes (`sc_method`, asynchronous Thread, and clocked Thread). SystemC control flow objects provide the groundwork on top of which complex user-defined control structures can be defined, e.g. hierarchical finite state machines (HF-sms) based on Harel statecharts [135],

- intra-module interaction, including computation, communication and synchronization performed by local processes, or remote control flow objects defined in other modules, using possibly dynamic thread managements, i.e. thread creation, interruption and abort; notice that asynchronous system modeling extensive random testing of synchronous systems must be based on modifying the deterministic, sequential process scheduler of the SystemC kernel [241] [287],

- standardized and user-defined inter-module point-to-point and multi-point communication channels and corresponding interfaces, e.g. peripheral, basic and advanced VCI [373], Amba bus (AHB, APB) [10], and ST Microelectronics' proprietary STBus (type1, type2 and type3) [334] [335]. These objects are built on top of SystemC in a straightforward manner using generic, usually IMC-based communication objects and protocols. As an example of communication modeling, our environment, called OCCN, is an open-source, SystemC-based library of objects with the necessary semantics for modeling efficiently on-chip communication networks [82]. Implementation details are provided in Section 5.3, and especially in the OCCN user manual available with the supplementary book CD.

System-level modeling is an essential ingredient of system design flow. Data and control flow abstraction of the system hardware and software components express not only functionality, but also performance application- and system- (or component-) specific characteristics that are necessary to identify system bottlenecks. For example, virtual channels are used to avoid both network and protocol deadlock, and also to improve performance and provide quality of service. While for software components it is usually the responsibility of the user to provide appropriate performance measurements, for hardware components and interfaces it is necessary to provide a statistical package that hides internal access to the modeling objects. This statistical package can be used for evaluating system performance characteristics. The statistical data may be analyzed using visualization software, e.g. the open source Grace tool, or dumped to a file for subsequent data processing, e.g. via an electronic spreadsheet or a specialized text editor.

Considering the hardware system modeling objects previously proposed, we observe that dynamic performance characteristics, e.g. latency, throughput, packet loss, hit ratio, resource utilization (e.g. buffer size), and possibly power consumption (switching activity), are definitely required from

- inter-module communication and synchronization objects, such as communication channels, and

- intra-module passive memory objects (Register, FIFO, LIFO, circular FIFO, Memory, Cache).

Since certain complex objects may be implemented in various ways, e.g. using control logic and static memory, different performance evaluation methods may be applied. Complex systems involve two general monitoring classes for extracting time-driven (instant) and event-driven (duration) statistics from system components.

We have described a general and rich system-level modeling and performance evaluation methodology that allows multiple levels of abstraction, enables hardware/software co-design, and provides interoperability between several state-of-the-art libraries and tools for long-term reuse opportunities, e.g. ISS, system and hardware description languages, and simulation interfaces. To provide reuse opportunities, many-to-many interoperability of today's highly complex and diversified toolkits and languages can be achieved through middleware, e.g. by inventing and standardizing a distributed SystemC-oriented Corba-like interface definition language [268]. In this case IPs selected from different providers can communicate, without exposing their internal structure or operations.

Conventional data recording, VCD tracing and source-code debugging are inadequate for performance monitoring and debugging complex, inherently parallel system models. Efficient design space exploration requires advanced, potentially overlapping system-level monitoring activities; notice that implementation issues relating to intrusiveness and synchronization are crucial.

- Generation refers to detecting events, and providing status and event reports containing monitoring traces (or histories) of system activity.

- Processing refers to functions that deal with monitoring information, such as filtering, merging of traces, correlation, analysis, validation, and model updating. These functions convert low-level monitoring data to the required format and level of detail.

- Presentation refers to displaying monitoring information in appropriate form.

- Dissemination concerns distribution of selected monitoring reports to system-level designers and external processing entities.

More details on monitoring classes (including an OCCN implementation) are provided in Section 5.3, and especially in the OCCN user manual in this book supplementary CD.

In addition to user-friendly text-based debugging and monitoring facilities, graphical visualization enhances this methodology by including dynamic and static GUIs [189] [287] [320] for

- interactive model design by importing ready-to-use basic and complex library modules, e.g. hierarchical FSMs, or reusable IP block or system components,

- simulation control, such as saving into files, starting, pausing and restarting simulation, displaying waveforms, and dumping or changing simulation and model parameters during initialization or runtime [65] [66], and

- platform-specific performance monitoring for metrics, such as simulation efficiency, and computation, and communication load.

System modeling for embedded software design can benefit from efficient parallel and distributed simulation frameworks [151] [268]. Distributed simulation performance is expected to be orders of magnitude better than existing commercial simulation tools. Furthermore, distributed simulation could support efficient time- and space-efficient modeling of large modular systems at different abstraction levels.

5.1.5 Design Space Exploration

In order to evaluate the vast number of complex architectural and technological alternatives the architect must be equipped with a highly-parameterized, user-friendly, and flexible design space exploration methodology. This methodology is used to construct an initial implementation from system requirements, mapping modules to appropriate system resources. This solution is subsequently refined through an iterative improvement strategy based on reliability, power, performance, and resource contention metrics. These metrics are obtained through domain- or application-specific performance evaluation using stochastic analysis models and tools and benchmarks for common applications, e.g. commercial networking or multimedia. The proposed solution provides values for all system parameters, including configuration options for sophisticated multiprocessing, multithreading, prefetching, and cache hierarchy components.

After generating an optimized mapping of behavior onto architecture, the designer may either manually decompose hardware components to the RTL level of abstraction or load system-level configuration parameters onto available behavioral synthesis tools, such as the Synopsys coreBuilder and coreConsultant. These tools integrate a preloaded library of configurable high-level (soft) IPs, e.g. ST Microelectronics' STbus [334] [335], VSIA's VCI [373], OCP [137], and generic interface prototypes. The target is to parameterize IP in order to generate automatically the most appropriate synthesis strategy. Now design flow can proceed normally with routing, placement, and optimization by interacting with tools, such as Physical Compiler, Chip Architect, and PrimeTime.

5.1.6 System-Level Design Tools

For design productivity to keep pace with technology, new high-level SoC design tools must provide accurate, yet flexible analysis to deal with early and partial design decisions and constraints. High-level tools and design environments can be categorized into system-level modeling, design space exploration (including simulation, performance analysis, power consumption estimation and reliability) and verification.

Although more recently C/C++ and SystemC use for hardware description and design flow has gained popularity, many design environments initially focused on co-design and co-verification techniques using multi-language descriptions, i.e. ISS for processors, HDL for hardware and C or similar languages for software. Such design tools include Polis [28], COSYMA [261], Ptolemy [207], IPSIM [81] and commercial tools, such as Cadence's VCC [73], CoWare [84], Synopsys' System Studio and Cocentric SystemC Compiler [346] [347] and Mentor Graphics' Seamless [305].

Most design tools operate at different levels of abstraction. Top-down design based on communication refinement is followed in most tools. Thus, we start from a functional description of the complete system, which is gradually refined into hardware IP ready for HDL simulation, floor planning and logic synthesis (e.g. embedded processors, memory, and hardware units), and software IP which is run on an embedded processor by linking to a selected RTOS configured with the appropriate interrupt handlers, counter/timer initialization and schedulers.

Synopsys System Studio was a design entry and modeling environment that utilized block and state diagramming techniques using graphical abstractions to represent concurrency, reactive control and dynamic switching concepts [346]. System Studio allowed the user to execute models either on a time-driven or an event-driven simulation engine. The Cocentric SystemC compiler integrates behavioral and RTL synthesis engines to transform SystemC descriptions into gate-level netlists, including support for resource sharing, pipelined operations, FSM generation, and a "MemWrap" graphical interface for mapping to specific types of memories [347]. In addition, "Bcview" provided a graphical cross-linking capability to source code.

ST Microelectronics' IPSIM is a design environment consisting of a SystemC-based extensive C++ modeling library, together with a simulation engine, runtime and test environment [81]. IPSIM's two-level approach communication refinement is similar to system- and application-level transactions in Cosy [51], which was based on concepts from VCC framework [73]. The system-level layer was responsible for selecting communication parameters and resolving latency-throughput tradeoffs, adopting eventually physical bus protocols. IPSIM design methodology has been applied to large-scale design projects in application domains, such as high speed networks.

Although all above tools have been withdrawn, more recently there is extensive research on multiprocessor system modelling and Multicore design tools. Thus, new tools start to appear in the market, e.g. targeting efficient application- and function-specific high-level synthesis, design space exploration or power estimation of multimedia or network processors. Several examples are provided below.

- The ST Microelectronics StepNP is a system-level SystemC-based design exploration platform for network processors [268]. It is a multithreaded architecture platform based on a high-level multiprocessor architecture simulation model, a network router application framework and a SoC interaction control, instrumentation, debugging and analysis toolkit.

- CoWare has developed ConvergenSC, a fast SystemC based system-level modeling and verification tool that enables rapid creation, simulation and analysis of transaction-level models for multiprocessor systems connected by complex on-chip buses [84].

- Unlike StepNP and ConvergenSC which focus on transaction-level communication, Benini et al. focus on multiprocessor architecture modelling based on ISS [32].

- Tensilica provides XTMP framework which can integrate multiple C-callable Xtensa configurable processors connected by customized interconnect modules [354]. Currently XTMP only supports functional simulation.

- A fast power-performance simulator for programmable processors and eventually on- and off-chip interconnection networks has been developed at Princeton University [278] [375] [396].

- Finally, the GRAPES framework is an integrated system-level co-design framework that supports cycle-level simulation of heterogeneous Multi-core systems, including debugging tools, performance, and power statistics collectors [236]. It features TLM-based bi-directional links, flexibility, modularity, time accuracy, and simulation speed, since it implements a light and fast simulation kernel. GRAPES loads and connects together as dynamic shared libraries C++ or SystemC models, such as ISS, memory, and bus components.

5.2 NoC Design Methodology and Tools

In the paradigm shift from device-centric to interconnect-centric application-specific Multicore, NoC design methodology and tools are vital to increase design productivity and effectively explore architectural, algorithmic and technological issues, such as topological properties, embedding quality for common communication patterns, performance, reliability and energy efficiency. Thus, NoC design shares many goals and constraints with SoC design, such as minimizing cost, energy and chip area, and maximizing concurrency and performance.

NoC design methodology focuses on obtaining a functionally correct OSI-like layered NoC topology that interconnects selected standard or custom IP resources, ranging from configurable hardware cores to complete multiprocessor array systems often selected using extensive design reuse principles, e.g. IP selection, qualification, customization phases. Thus, NoC design tools try to explore static or dynamic mapping and scheduling of application functionality onto the heterogeneous NoC platform to form a concrete system; this includes resource allocation, architecture simulation, optimization and verification of QoS criteria, including performance and power consumption.

In addition, NoC design tools often generate SystemC or HDL code (through high-level synthesis) that can be used to estimate area and power consumption or provide adapters for simple integration of standardized IP blocks. Currently available NoC design tools include the Silistix ChainWorks NoC design tool and the open-source On-Chip Communication Network (OCCN) framework.

In addition to NoC emulation using FPGAs [120], several research groups have proposed simulation tools based on cycle-accurate simulation in VHDL, SystemC or combinations to specify, simulate, analyze and generate NoCs at different levels of abstraction. Although FPGA-based emulation takes only a few simulation seconds (compared to hours) for processing millions of cycles of complex NoC behavior, simulation methods offer maximum design flexibility. Next, we outline several simulation tools.

5.2.1   NoC Simulation Tools

OPNET is a generic telecommunication network simulator which can be conveniently adapted to hierarchically model and simulate on-chip interconnections architectures based on different traffic scenarios [260]. OPNET environment has been used in modeling QNoC. However, certain limitations exist, such as support for asynchronous communications only [390].

The network-on-chip interconnect calculator tool (NoCIC) predicts performance and power for NoC structures using a set of design parameters simulated with low-level SPICE tools [369].

The NS-2 simulator provides support to define the network topology, communication protocol, routing algorithm and traffic [341]. NS-2 provides traces for simulation results interpretation and a message flow graphical visualization tool.

NoCSim is a NoC simulator based on processing elements. These elements respond to network packets sent to them by their associated switch. NoCSim generates traffic using constant bit rate and random Poisson distribution [386].

A modular system-level performance modeling and design space exploration framework for on-chip interconnects, such as dedicated shared bus and point-to-point network topologies, has been proposed by Kogel [188].

The Hermes NoC infrastructure automatically generates packet-switched NoCs for mesh-like topologies [237]. It supports three OSI-like layers: physical for the communication links, data link for reliable communication exchange during handshake, and network implementing packet switching. The Hermes IP to switch interface may use either a native interface or the OCP standard point-to-point master-slave interface for enhanced design time, reusability and connectivity to available compliant IP cores [255]. With this protocol, master (typically a processor) initiates commands by issuing read and write transactions, while slave (typically a memory) responds by receiving data and possibly sending data back to the master. The Maia tool is an interesting graphical framework enabling user-driven Hermes NoC parameterization, automated NoC (and core network interface) code generation in RTL or SystemC, and statistical NoC traffic generation and seamless analysis of architecture, application and technology parameters [229]. It enables automated selection of several configuration parameters flow control (asynchronous handshake vs. credit-based), number of virtual channels and scheduling (round robin vs. priority), mesh size, packet and flit size, buffer size (input or central buffering), routing algorithm (xy-routing or Turn model for adaptive routing) and SystemC-based test benches. Although Maia initially supports only best effort, i.e. transmitted packets can be arbitrarily delayed by the network, in the future it could provide guaranteed throughput via circuit switching and/or virtual channels.

Alike Maia, the NoCGEN tool provides NoC parameterization and mixed VHDL/SystemC implementation and simulation using a template router that supports several interconnection networks. It creates top-level VHDL description of the NoC and a mixed VHDL/SystemC test bench environment for simulation and synthesis [62]. NoCGen configures wormhole-based output-queued 2-d mesh routers with a different number of ports, routing algorithms, data widths and buffer depths. Besides, it presents a mixed SystemC VHDL simulation environment.

Another NoC generation and simulation environment for the Philips Æthereal NoC is based on parameterized XML-based configuration files for topology, mapping and master-slave connection which implicitly describe the network interface, router, and application traffic components [270]. It is also possible to describe the IP blocks used to generate a SoC. A NoC is eventually simulated in SystemC and RTL VHDL and evaluated in terms of cost and performance by sweeping over different parameters. An independent performance verification tool verifies analytically that the NoC meets application performance requirements.

Apart from NoC design-specific tools discussed in Chapter 2, e.g. xpipes and Arteris configurable NoC IP, other state-of-the-art NoC design tools include the Silistix ChainWorks NoC design tool and the open-source On-Chip Communication Network (OCCN) framework.

Silistix, an initially venture funded spinout from the University of Manchester, develops IP libraries and software tools to automate NoC design space exploration [27].

Silistix CHAINworks is a suite of EDA design tools that integrate asynchronous self-timed synthesis technology for designing an application-specific mix of bus and crossbar topologies that meets performance, area and power tradeoffs [317]. GALS methodology, and in particular asynchronous handshake protocols based on delay-insensitive data encodings, enable NoCs to operate at a much higher cycle rate, fully-decoupled from processor and memory subsystems. This results in greater design flexibility, reduced time-to-market through rapid SoC timing closure during system integration, and improved reliability, performance and power dissipation. In particular, through GALS methodology CHAINworks supports multiple unrelated clock domains, thus eliminating clock distribution buffers and drivers and allowing SoC cores to operate at their natural operating frequency rather than worst-case imposed by a system clock tree.

CHAINworks uses adapters to interface existing peripherals and synchronous bus architectures, such as OCP 2.0, IBM CoreConnect, and AMBA AHB, APB and AXI promoting block- and system-level reuse.

Silistix CHAINworks allows designers to create an application optimized interconnect that balances performance, area, and power consumption to fit system requirements. CHAINworks tools include: CHAINdesigner, CHAINcompiler, and CHAINlibrary.

CHAINdesigner is a graphical design tool used to specify topologies and attributes of self-timed on-chip interconnects for SoC. The tool allows referencing existing Verilog modules for modular, plug-and-play SoC design. CHAINdesigner generates behavioral Verilog and SystemC NoC models, as well as simulation test benches with the appropriate network topology, link widths, and pipeline repeaters to meet requirements at a minimum cost.

CHAINcompiler synthesizes the constrained netlist generated by CHAINdesigner using highly parameterized circuit components from the CHAINlibrary (mapped to standard cells) to produce the structural Verilog netlist suitable for input into conventional logic synthesis tools, such as Synopsys' Design Compiler for standard cell SoC and ASIC design. In addition, CHAINcompiler provides test patterns for manufacturing, as well as scripts for static timing analysis and placement and routing using existing tools.

Another design environment for NoC modeling and simulation is OCCN (open-source On-Chip Communication Network) [80] [82]. This framework is based on an object-oriented C++ library, built on top of SystemC. It enables NoC creation in different abstraction levels, such as transaction-level modeling. Besides these features, OCCN allows protocol refinement, design exploration, and NoC components design and verification based on prototype communication APIs. OCCN is briefly discussed next, while library details and a user manual are provided in the supplementary book CD.

5.3 The On-Chip Communication Network (OCCN)

This Section focuses on OCCN, the only open-source SystemC-based NoC modeling and design space exploration environment currently available. On-Chip Communication Network (OCCN) originated as a "community of practice" research activity at AST-Grenoble among an informal community of people interested in sharing experiences and ideas on a new breakthrough technology for next generation System-on-Chip (SoC) interconnects. Within this research activity, based on community experience, state-of-the-art NoC design methodology and best practices have been developed. To enrich this learning process and improve NoC design methodology research not only within ST Microelectronics but also externally, an open-source project was established to enable knowledge sharing and competence on innovative NoC design methodology tools in a horizontal learning environment, not only within ST Microelectronics Divisions but also externally through contacts with international research community, prestigious Academia and industry.

From a technical viewpoint, OCCN provides a flexible, highly-parameterized, state-of-the-art framework consisting of an open-source, GNU GPL library built on top of SystemC, aiming at complex NoC modeling and high-level performance modeling at various levels of abstraction, thus enabling efficient system-level simulation and design space exploration of a complex design space. OCCN focuses on NoC architecture and topology issues, e.g. router–arity, packet/flit size, flow control, buffer management, packet routing and virtual channel allocation strategies at the router and network interface. OCCN's ultimate goal is bringing an original methodology into the design of complex interconnect models. To achieve very high simulation speed, typical TLMs of interconnects are either purely functional (without precise timing information) or clock-approximate. These kinds of models provide the SoC software architect with virtual platforms encompassing ISS, DMA, and memory blocks from early in the design cycle. On top of these platforms, complex SoC software and debuggers can be run far before RTL system models are ready or integration is completed. This approach is severely limited for interconnect modeling where clock accuracy is fundamental in avoiding common pitfalls, such as overcommitment, and providing correct network design, reduced contention, optimized arbitration and flow control. OCCN fosters the development of clock-accurate TLMs providing well-defined blocking and non-blocking communication APIs, on top of which complex NoC models can be easily built.

OCCN increases the productivity of developing communication driver models through the definition of a universal Application Programming Interface (API). This API provides a new design pattern that enables creation and reuse of executable transaction level models (TLMs) across a variety of SystemC-based environments and simulation platforms. It also addresses model portability, simulation platform independence, interoperability, and high-level performance modeling issues.

OCCN-based models have already been used by Academia and Industry, such as ST Microelectronics, for developing and exploring a new design methodology for on-chip communication networks. In the past, this methodology enabled the design of next generation networking and home gateway applications, and complex on-chip communication networks, such as the ST Microelectronics proprietary bus STBus, a real product found today in almost any ST chip for HDTV, set top box, mobile phone, or digital satellite decoder [301] [302].

Next in Sections 5.3.1 to 5.3.3, we outline the OCCN API, focusing on the establishment of inter-module communication refinement through a layered approach based on two SystemC-based modeling objects: the Protocol Data Unit (Pdu) and the MasterPort and SlavePort interface. In Section 5.3.4, we provide implementation details of OCCN communication channel design methodology, focusing on point-to-point (`StdChannel`) and multi-point (`StdBus`) channels. We also explain a simple router component for NoC modeling (`StdRouter`). The advanced user may be able to exploit this methodology for implementing high-level, specialized on-chip communication protocols, increasing NoC design productivity. Finally, in Section 5.3.4.1, we present a transmitter/receiver case study illustrating inter-module communication refinement, high-level system performance modeling based on generic OCCN statistical classes and future extensions at the user level.

5.3.1   The OCCN Methodology

As in all system development methodologies, any SoC object oriented modeling would consist of a modeling language, modeling heuristics and a methodology [307]. Modeling heuristics are informal guidelines specifying how the language constructs are used in the modeling process. Thus, the OCCN methodology focuses on modeling complex on-chip communication network by providing a flexible, open-source, object-oriented C++-based library built on top of SystemC. System architects may use this methodology to explore NoC performance tradeoffs for examining different on-chip communication architectures (OCCA) implementations.

OCCN provides several important modeling features.

- Object-oriented design concepts, fully exploiting advantages of this software development paradigm.

- Optimized design based on system modularity, refinement of communication protocols, and IP reuse principles. Notice that even if we completely change the internal data representation and implementation semantics of a particular system module (or communication channel), while keeping a similar external interface, users can continue to use the module in the same way.

- Reduced model development time and improved simulation speed through powerful C++ classes.

- System-level debugging using a seamless approach, i.e. the core debugger is able to send detailed requests to the model, e.g. dump memory, or insert breakpoint.

- Plug-and-play integration and exchange of models with system-level tools supporting SystemC, such as System Studio (Synopsys), NC-Sim (Cadence), and CoWare, making SystemC model reuse a reality.

- Efficient simulation using direct linking with standard, nonproprietary SystemC versions.

- Early design space exploration for defining the merits of new ideas in OCCA models, including high-level system performance modeling.

OCCN models have two main characteristics.

- First, they do not use expensive SystemC signals at module interfaces that heavily burden the SystemC scheduler with several request/update calls for each signal update. OCCN efficiently models channel signals using plain variables; signal consistency is guaranteed through careful scheduling of function calls.

- Secondly, OCCN models require far less SystemC control than typical RTL models, since scheduling is performed statically and does not rely much on the SystemC kernel. For example, typical OCCN models provide a single SystemC thread or method per network channel, considerably reducing SystemC scheduler overhead and improving simulation speed by a significant factor.

Alike OSI layering, OCCN methodology for NoC establishes a conceptual model for inter-module communication based on layering, with each layer translating transaction requests to a lower-level communication protocol. As shown in Figure 5.1, OCCN methodology defines three distinct OCCN layers. The lowest layer provided by OCCN, called NoC communication layer, implements one or more consecutive OSI layers starting by abstracting first the Physical layer. For example, the STNoC router communication layer abstracts the physical and data link layers. On top of the OCCN protocol stack, the user-defined application layer maps directly to the application layer of the OSI stack. Sandwiched between the application and NoC communication layers lies the adaptation layer that maps to one or more middle layers of the OSI protocol stack, including software and hardware adaptation components. The aim of this layer is to provide, through inter-dependent entities called communication drivers, the necessary computation, communication, and synchronization library functions and services that allow the application to run.

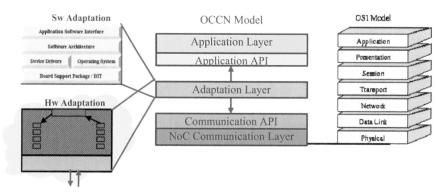

FIGURE 5.1:   OSI-like OCCN layering model with APIs

Although the adaptation layer is usually user-defined, it utilizes functions defined within the OCCN communication API.

An implementation of an adaptation layer includes software and hardware components, as shown in the left part of Figure 5.1. A typical software adaptation layer includes several sub-layers. The lowest sub-layer is usually represented by the board support package (BSP) and built in tests (BIT). The BSP allows all other software, including the Operating System (OS), to be loaded into memory and start executing, while BIT detects and reports hardware errors. On top of this sub-layer we have the OS and device drivers. The OS is responsible for overall software management, involving key algorithms, such as job scheduling, multitasking, memory sharing, I/O interrupt handling, and error and status reporting. Device drivers manage communication with external devices, thus supporting the application software. Finally, the software architecture sub-layer provides execution control, data or message management, error handling, and various support services to the application software.

The OCCN conceptual model defines two APIs.

- The OCCN communication API provides a simple and generic interface that greatly simplifies the task of implementing various layers of communication drivers at different level of design abstraction. The API is based on generic modeling features, such as IP component reuse and separation between behavior and communication. It also hides architectural issues related to the particular on-chip communication protocol and interconnection topology, e.g. simple point-to-point channel vs. complex, multilevel NoC topology supporting split transactions, and QoS in higher communication layers, thus making internal model behavior module-specific. The OCCN communication API is based on a message-passing paradigm providing a small, powerful set of methods for inter-module data exchange and synchronization of module execution. This paradigm forms the basis of the OCCN methodology, enhancing portability and reusability of all models using this API.

- The application API forms a boundary between the application and adaptation layers. This API specifies the necessary methods through which the application can request and use services of the adaptation layer, and the adaptation layer can provide these services to the application.

The OCCN implementation for inter-module communication layering uses generic SystemC methodology, e.g. a SystemC port is seen as a service access point (SAP), with the OCCN API defining its service. Applying the OCCN conceptual model to SystemC, we have the following mapping.

- The NoC communication layer is implemented as a set of C++ classes derived from the SystemC `sc_channel` class. The communication channel establishes the transfer of messages among different ports according to the protocol stack supported by a specific NoC.

- The communication API is implemented as a specialization of the `sc_port` SystemC object. This API provides the required buffers for inter-module communication and synchronization and supports an extended message passing (or possibly shared memory) paradigm for mapping to any NoC.

- The adaptation layer translates inter-module transaction requests coming from the application API to the communication API. This layer is based on port specialization built on top of the communication API. For example, the communication driver for an application that produces messages with variable length may implement segmentation, thus adapting the output of the application to the input of the channel.

The fundamental components of the OCCN API are the Protocol Data Unit (Pdu), the MasterPort and SlavePort interface, and high-level system performance modeling. These components are described in the following sections.

5.3.2 The Protocol Data Unit (PDU)

In OCCN, inter-module communication is based on channels implementing well-specified protocols by defining rules (semantics) and types (syntax) for sending and receiving protocol data units (or Pdus, according to OSI terminology). In general, Pdus may represent bits, tokens, cells, frames, or messages in a computer network, signals in an on-chip network, or jobs in a queuing network. Thus, Pdus are a fundamental ingredient for implementing inter-module (or inter-PE) communication using arbitrarily complex data structures.

A Pdu is essentially the optimized, smallest part of a message that can be independently routed through the network. Messages are variable in length and usually consist of several Pdus composed of header control and data fields.

Furthermore, within the OCCN channel, several important routing issues involving Pdus must be explored. Thus, OCCN defines several functions that support simple and efficient interface modeling, such as adding or stripping headers from Pdus, copying Pdus, error recovery, flow control, segmentation and re-assembly procedures for adapting to physical link bandwidth, service access point selection, and connection management. Furthermore, the Pdu specifies the format of the header and data fields, the way that bit patterns must be interpreted, and any processing to be performed (usually on stored control information) at the sink, source or intermediate network nodes.

More details of the Pdu class with C++ examples with different types of packets, including header and/or body (data field and trailer), are illustrated in the OCCN user manual.

5.3.3   The MasterPort and SlavePort API

In this Section, we describe the transmission/reception interface of the OCCN API. The OCCN API provides a message-passing interface, with send and receive primitives for point-to-point and multi-point communication. If the Pdu structure is determined according to a specific OCCA model, the same base functions are required for transmitting the Pdu through almost any OCCA. A great effort is dedicated to define this interface as a reduced subset of functions providing users with a complete and powerful semantic. In this manner, we can achieve model reuse and inter-module communication protocol refinement through a generic OCCA access interface.

Message passing systems can emulate efficiently different communication paradigms, including shared memory, remote procedure calls, or Ada-like rendezvous. The semantic flavor of message passing primitives may specify when buffers (or data structures) can be reused without compromising correctness, or it may reflect completion semantics, i.e. when control is returned to the user process that issued the send or receive. There are two major point-to-point message passing primitives: synchronous and asynchronous. Synchronous primitives are based on acknowledgments, while asynchronous primitives usually deposit and remove messages directly to/from user application leading to simpler programming and less synchronization latency, but also less robustness and scalability since available buffer space may be too short or easily wasted. Within the class of asynchronous point-to-point communications, there are also two other major principles: blocking and non-blocking. While non-blocking operations allocate additional system buffers to enable the calling process to continue execution, e.g. by scheduling another thread, blocking operations suspend execution until a timeout occurs, or the message is either submitted to the network or becomes available in the application buffer.

Synchronous blocking send/receive primitives offer the simplest semantics for the programmer, since they involve a handshake (rendezvous) between sender and receiver.

- A synchronous send busy waits (or suspends temporarily) until a matching receive is posted and receive operation has started. Thus, the completion of a synchronous send guarantees that the message has been successfully received, and that all associated application data structures and buffers can be reused. A synchronous send is usually implemented in three steps.

  - First, the sender sends a request-to-send message.
  - Then, the receiver stores this request.
  - Finally, when a matching receive is posted, the receiver sends back a permission-to-send message, so that the sender may send the packet.

- Similarly, a synchronous receive primitive busy waits (or suspends temporarily) until there is a message to read.

With asynchronous blocking operations we avoid polling, since we know exactly when the message is sent and received. The communication semantics for point-to-point asynchronous blocking primitives are defined as follows.

- The blocking send busy waits (or suspends temporarily) until the packet is safely stored in the receive buffer (if the matching receive has already been posted) or in a temporary system buffer (message in care of the system). Thus, the sender may overwrite the source data structure or application buffer after the blocking send operation returns. Compared to a synchronous send, this allows the sending process to resume sooner, but the return of control does not guarantee that the message will actually be delivered to the appropriate process. Obtaining such a guarantee would require additional handshaking.

- The blocking receive busy waits (or suspends temporarily) until the requested message is available in the application buffer. Only after the message is received, the next receiver instruction is executed. Unlike a synchronous receive, a blocking receive does not send an acknowledgment to the sender.

For point-to-point asynchronous non-blocking send/receive primitives, we define the following semantics.

- A non-blocking send initiates the send operation, but does not complete it. The send returns control to the user process before the message is copied out of the send buffer. Thus, data transfer out of the sender memory may proceed concurrently with computations performed by the sender after the send is initiated and before it is completed. A separate send completion function, implemented by accessing (probing) a system communication object via a handle, is needed to complete the communication, i.e. for the user to check that the data have been copied out of the send buffer, so that the application data structures and buffers may be reused; notice that blocking send (receive) is equivalent to non-blocking send (respectively, receive) immediately followed by a blocking send (respectively, receive) completion function call. These functions either block until the desired state is observed or return control immediately reporting the current send status.

- Similarly, a non-blocking receive initiates the receive operation, but does not complete it. The call will return before a message is stored at the receive buffer. Thus, data transfer into receiver memory may proceed concurrently with computations performed after receive is initiated and before it is completed. A separate receive completion function, implemented by accessing (probing) a system communication object via a handle, is needed to complete the receive operation, i.e. for the user to verify that data have been copied into the application buffer. These probes either block until the desired state is observed or return control immediately reporting the current receive status.

Although there are literally hundreds of ways to combine synchronous, asynchronous blocking, and asynchronous non-blocking send/receive function calls, semantics becomes very complex. The precise type of send/receive semantics to implement depends on how the program uses its data structures, and how much we want to optimize performance over ease of programming and portability to systems with different semantics. For example, asynchronous sends alleviate the deadlock problem due to missing receives, since processes may proceed past the send to the receive process. However, for non-blocking asynchronous receive, we use a probe before actually using the received data.

In order to increase modeling efficiency, the OCCN MasterPort and SlavePort API is based only on synchronous blocking send/receive and asynchronous blocking send primitives. The proposed methods support synchronous and asynchronous communication, based on either a synchronous (cycle-based) or asynchronous (event-driven) OCCA model.

FIGURE 5.2:   MasterPort and SlavePort derived from sc_port

As shown in Figure 5.2, this API is implemented using two specializations of standard SystemC **sc_port< ...>**, called **MasterPort< ...>** and

SlavePort< ...>. The name Master/Slave is given just for commodity reasons, there is no relationship with the Master/Slave library provided in SystemC, or RPC concepts introduced earlier by CoWare [44]; this encompasses the limitation that the Master port is always connected to a Slave one. In our case, the name Master is associated to the entity, which is responsible to start the communication. Master and Slave ports are defined as templates of the outgoing and incoming Pdu. In most cases, the outgoing (incoming) Pdu for the Master port is the same as the incoming (respectively, outgoing) Pdu for the Slave port.

Hereafter, a list of OCCN MasterPort and SlavePort API functions is provided.

- `void send(Pdu< ...>* p, sc_time& time_out=-1, bool& sent)` implements synchronous blocking send. Thus, the sender will deliver the Pdu p, only if the channel is free, the destination process is ready to receive, and the user-defined `time_out` value has not expired. Otherwise, the sender is blocked and the Pdu is dispatched. While the channel is busy (or the destination process is not ready to receive), and the `time_out` value has not expired, waiting sender tasks compete to acquire the channel using a FIFO priority scheme. Upon function exit, the boolean flag `sent` returns false, if and only if the `time_out` value has expired before sending the Pdu; this event signifies that the Pdu has not been sent.

- `void asend(Pdu< ...>* p,sc_time& time_out=-1, bool& dispatch)` implements asynchronous blocking send. Thus, if the channel is free and the user-defined `time_out` value has not expired, then the sender will dispatch the Pdu p whether or not the destination process is ready to receive it. While the channel is busy, and the user-defined `time_out` value has not expired, waiting sender tasks compete to acquire the channel using a FIFO priority scheme. In this case, the boolean flag `dispatch` returns false, if and only if the `time_out` value has expired before sending the Pdu; this event signifies Pdu loss.

- The OCCN API implements a synchronous blocking receive using a pair of functions: `receive` and `reply`.

  - `Pdu< ...>* receive(sc_time& time_out=-1, bool& received)` implements synchronous blocking receive. Thus, the receiver is blocked until it receives a Pdu, or until a user-defined `time_out` has expired. In the latter case, the boolean flag `received` returns false, while the Pdu value is undefined.

  - `void reply(uint delay=0)` or `void reply(sc_time delay)` indicates that the receiver process completes the transaction in a dynamically variable delay `time`, expressed as a number of bus cycles or as absolute time (`sc_time`). Return from reply ensures that the channel send/receive operation is completed and that the receiver is synchronized with the sender. The following code is used for receiving a Pdu.

```
sc_time timeout =  ....;
bool received;
// in is a SlavePort
Pdu< ...> *msg = in.receive(timeout, received);
if (!received)
// timeout expired:  received Pdu not valid
else
// elaboration on valid received Pdu
reply(); // synchronizing after 0 cycles
```

Notice that when the delay of a transaction is measured in terms of bus cycles, OCCN assumes that the channel is the only one to have knowledge of the clock, allowing asynchronous processes to be connected to synchronous clocked communication media. In both cases the latency of reply can be fixed or dynamically calculated after the receive, e.g. as a function of the received Pdu.

An example of fixed receive latency delay is provided below.
```
sc_time latency(2, SC_NS);
msg=in.receive(); // msg available from SlavePort in
addr=occn_hdr(*msg, addr);
seq=occn_hdr(*msg, seq); // managing the payload
reply(latency);//receiver synchronized with transmitter
```

An example of dynamic delay is given below.
```
uint latency=5; // latency expressed in bus cycles
msg=in.receive(); // obtain msg with a dynamic delay
addr=occn_hdr(*msg, addr);
seq=occn_hdr(*msg, seq);
latency=delay_function(msg); // dynamic latency (on msg)
reply(latency);//receiver synchronized with transmitter
```

Furthermore, notice that a missing reply to a synchronous send could cause a deadlock, unless a sender timeout value is provided. In the latter case, we allow that the Pdu associated with the missing acknowledgment is lost. Notice that a received Pdu is also lost, if it is not accessed before the corresponding reply.

Sometimes tasks may need to check, enable, or disable Pdu transmission or reception, or extract the exact time(s) that a particular communication message arrived. These functions enable optimized modeling paradigms that are accomplished using appropriate OCCN channel setup and control functions. A detailed description of these functions falls outside the scope of this document [80].

Another OCCN feature is protocol in-lining, i.e. the low-level protocol necessary to interface a specific OCCA is automatically generated using the standard template feature available in C++ enabled by user-defined data structures. This implies that the user does not have to write low-level communication protocols already provided by OCCN, thus making instantiation and debugging easier, and resulting in significantly reduced modeling effort.

5.3.4 OCCN Channel Design Methodology

Communication channels are responsible for inter-module communication, transferring both signals and data according to a given communication protocol (called channel interface). In general, a channel interface may model both direct, point-to-point communication, as well as multi-access channels, such as crossbar, bus, multistage or multi-computer network, thus forming a complex network-on-chip.

The OCCN library currently provides three channel interfaces: `StdChannel` that deals with bi-directional, point-to-point communication, `StdBus` that deals with traditional, simplified, bus-based communication, and `StdRouter` that models on-chip indirect connections among communication routers, such as those required for implementing crossbars, multistage and direct networks featuring a number of simultaneous connections among communication nodes. However, implementation of the basic OCCN communication channels provides a general SystemC-based inter-module communication design methodology that allows the design of other specialized, user-defined bus channels. For example, Figure 5.3 illustrates class inheritance for OCCN classes implementing point-to-point (`StdChannel`) and multi-point (`StdBus`) inter-module communication models. OCCN classes are shown in white, while related SystemC classes are shown in grey.

From Figure 5.3, notice that OCCN channels are defined as template objects of a pair of

- user-defined Pdu classes, such as classes `PduO` and `PduI` of `StdChannel`,

- channel-specific Pdu classes, or

- mixed Pdu classes, such as `Pdu<StdBusMasterCtrl, Data, Size>` and `Pdu<StdBusSlaveCtrl, Data, Size>` of `StdBus`, where only `Data` and `Size` are user-defined.

Thus, in the common case of Master/Slave communications, each channel can be parameterized through a set of user-defined or channel-specific signals and data coming out of the Master (and going to the Slave), and another set of signals and data coming out of the Slave (and going to the Master).

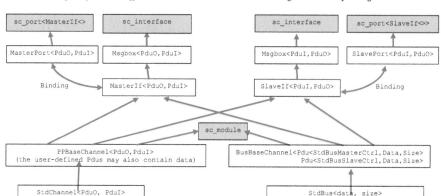

FIGURE 5.3: Class inheritance for point-to-point (`StdChannel`) and multi-point (`StdBus`) channels

These signals define the Module interface, and may include data, address and operation code fields in the Pdu header.

Next, we proceed to briefly outline implementation issues, concentrating on the operation and use of existing OCCN communication channels. Operation of `StdBus`, `StdChannel` and `StdRouter` is based on the `send/asend` and `receive/reply` routines implemented by the module ports. Implementation of these functions is explained, assuming a bi-directional `module A` to `module B` transfer. These transfers are similar to the ones in `StdChannel` and `StdBus`. While in Figure 5.4 we concentrate on synchronous send communications, in Figure 5.5 we concentrate on asynchronous asend communications. Notice that these figures do not show implementation of timeout management. Actually, timeouts are made preemptive by additional testing and implementation of special `cancel_sending` functions that directly translate into channel reset management.

In addition to the previously discussed send/asend and receive/reply functions, `backdoor_read(size, addr, buffer)` (and `backdoor_write`) functions allow access to any Slave outside of simulation scope, i.e. simulation time is not advanced and the context is not changed. These functions are useful for loading programs, initializing data, or for debugging, i.e. setting breakpoints or dumps. However, these module functions are first bound to the corresponding port using `port->back_door_read_register(&backdoor_read, this)` and also `port->back_door_write_register(&backdoor_write, this)`.

5.3.4.1 OCCN API and Library Components

OCCN ports and interfaces are integrated as a superset of basic SystemC ports and interfaces (`MasterPort`, `SlavePort`, `MasterIf`, `SlaveIf`) in order to define environment-specific access primitives as a reduced subset of functions for improving model portability.

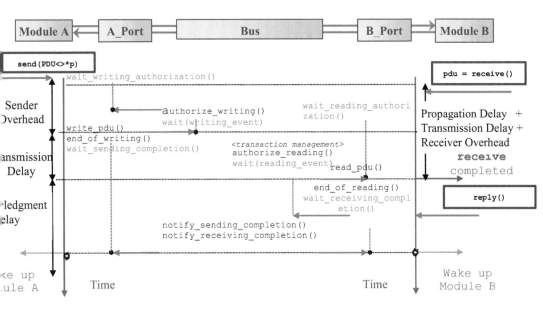

FIGURE 5.4: Synchronous communication using send and receive/reply

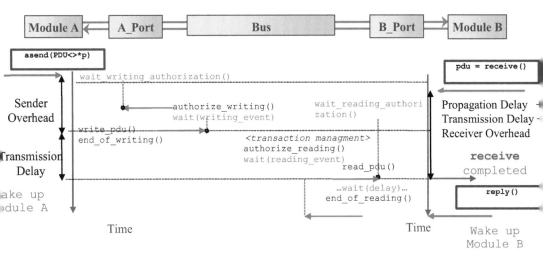

FIGURE 5.5: Synchronous communication asend and receive/reply

The `BusBaseChannel` class constitutes the main building block from which complex models of busses can be built. It contains functions useful at port binding to create dynamically Master and Slave Interfaces.

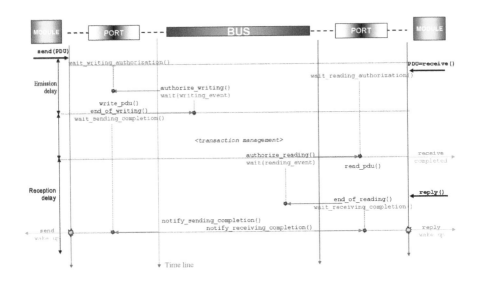

FIGURE 5.6: Send/Receive protocol example

The message-passing paradigm with send and receive primitives is used for sending and receiving Pdus through ports of point-to-point and multi-point communication channels. Figure 5.6 shows a simple schematic behavior (data flow) of send and receive primitives. Internal OCCN communication protocol control mechanisms and data flags, such as `notify_sending_completion()`, are used by NoC model developers to implement inter-module communication channels and handle transaction scheduling and model parameterization efficiently. Unlike SystemC control signals which invoke the kernel, OCCN control signals that control events in blocking (`send()`) and nonblocking (`asend()`) interfaces are implemented as plain C++ variables, thus achieving much higher computation speed than RTL simulation. However, notice that proper sampling of signals is necessary since variables do not guarantee thread order consistency in that state variables changes take effect immediately, not after a SystemC delta cycle.

In the following subsections, the functionality of OCCN API components is explained.

The increasing need for standard interfaces implies use of standard ports that also speed up development of common wrappers.

Since NoC protocol management is performed through simple interface method calls, e.g. send primitive to transfer a Pdu, and dynamic model is performed at the bit- and cycle-accurate abstraction level (as in RTL simulation), this type of modeling is essentially bit- and cycle-accurate transaction level modeling.

5.3.4.2 Master/Slave Interfaces

Master/Slave interfaces define the NoC protocol either directly, i.e. through function definitions, or indirectly, i.e. through virtual pointer dereferencing represent the place where the protocol is defined. OCCN developers made a design decision where the interface clearly draws a line between functions available to the models users and those available to models developers.

The Master/Slave Interface is mostly a container for the Msgbox object with extra utility functions for debugging and direct access (i.e. backdoor access) which skips channel arbitration. The Msgbox object defines several useful functions as shown below.

```
template <class WPdu, class RPdu>
class Msgbox : public sc_interface
{
public:

  // constructors
  Msgbox();
   Msgbox();

  // Access methods typically if Msgbox is used as master
  // module side
  N_int wait_write_authorization();
  N_int wait_write_authorization(sc_time& time_out);
  N_int ask_write_authorization();
  N_uint write_pdu(WPdu& _ref);
  void end_of_writing();
  N_int wait_sending_completion();
  N_int wait_sending_completion(sc_time& time_out);
  void cancel_sending();

  // channel side
  void authorize_writing();
  bool is_writing_completed();
  void assign_writing_event(sc_event*);
  void enable_writing_event();
  void disable_writing_event();
  void notify_sending_completion();
  void assign_sending_cancel_event(sc_event*);
```

```
bool is_sending_cancelled();
void reset_sending_cancel();
void reset_writing_access();

// Access methods typically if Msgbox is used as slave
// module side
N_int wait_read_authorization();
N_int wait_read_authorization(sc_time& time_out);
N_int ask_read_authorization();
N_uint read_pdu(RPdu* _ref);
void end_of_reading();
N_int wait_receiving_completion();
void cancel_receiving();

// channel side
void authorize_reading();
bool is_reading_completed();
void assign_reading_event(sc_event*);
void enable_reading_event();
void disable_reading_event();
void notify_receiving_completion();
void enable_receiving_completion_event();
void disable_receiving_completion_event();
void assign_receiving_cancel_event(sc_event*);
bool is_receiving_cancelled();
void reset_receiving_cancel();
void reset_reading_access();

// data/ctrl members access
WPdu* get_write_pdu_ptr();
RPdu* get_read_pdu_ptr();
void set_write_pdu_ptr(WPdu*); // shouldn't be used by user
void set_read_pdu_ptr(RPdu*); // shouldn't be used by user
};
```

The Msgbox object implements interface functionality by properly scheduling functions that manage communication behaviors, including consistent, statically scheduled, parallel or pipelined cycle-accurate TLM models, e.g. separately for virtual circuits and for input, switching, output stages. The code explicitly defines the functions available to access methods, such as send or receive, most visible only to the model developer. Access functions, such as wait_write_authorization(), are used by primitives, such as send, to ask channel control and to transfer data. Control functions, such as authorize_writing(), are used explicitly by model developers to control flow of data and protocol timing.

Let us suppose that a master module, possibly a Network interface (NIC) which contains a `MasterPort`, has just called a send primitive to send a Pdu to the downstream router (which contains and implements a Master Interface, so a `Msgbox`) as shown in Figure 5.7. The send primitive calls four functions: `wait_write_authorization()`, `write_pdu()`, `end_of_writing()` and `wait_sending_completion()` in this order. Next, we describe OCCN inner-workings by describing these functions step by step.

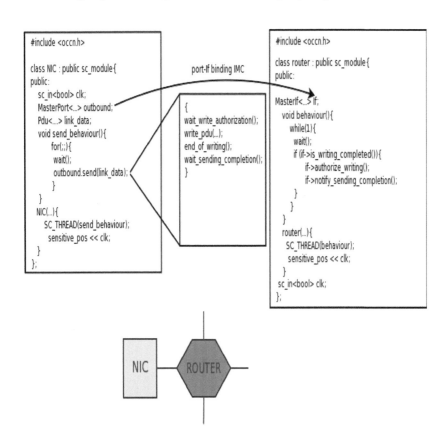

FIGURE 5.7: Example of protocol implementation using `MsgBox`

The `wait_write_authorization()` function tries to lock a mutex object. This function may block, when the mutex is already locked, allowing just a single thread to call it and access the channel at a given time. The mutex state is controlled by the router `behaviour()` thread (through utility functions, such as `authorize_writing()`), that can unlock the mutex, thus providing nonblocking behavior when the protocol allows it, see Figure 5.7).

The `write_pdu` function is an inline function that copies the Pdu in `MsgBox` internal storage.

The `end_of_writing()` function sets a flag used by (router) arbitration to check the send primitive completion, through `is_writing_completed()` function.

The function `wait_sending_completion()` is a blocking function that manages multiple blocking conditions on a single send call.

The `notify_sending_completion()` function releases the event on which the send call is blocked.

All these functions provide a simple but powerful set of primitives to manage any given point-to-point on-chip link protocol in a time- or event-driven manner by providing blocking and release conditions, and flags to avoid polling. In a real-world router, the router checks control fields, e.g. the Pdu to guarantee point-to-point protocol correctness, relating to bit- and cycle-accuracy.

5.3.4.3 Point-to-Point `StdChannel`

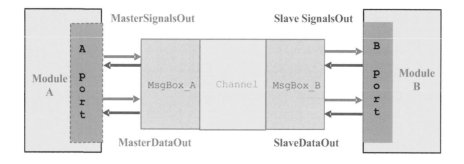

FIGURE 5.8: Point-to-point, inter-module communication with `StdChannel`

As shown in Figure 5.8, the `StdChannel` (standard channel interface) implements point-to-point communication between `module A` and `module B`. The channel allows for synchronous, bi-directional exchange of two Pdu data structures between the two modules. Both Pdu structures (called incoming and outgoing Pdus when referring to a single module) contain user-defined control signals and data, implemented separately and independently for each module. The bi-directional, point-to-point connection between each module and the channel is realized using two interfaces (`A_Port` and `B_Port`) located in the corresponding modules. These interfaces are compatible but not necessarily the same, e.g. signals and data handled by the ports may be different. Each interface is attached to a Message Box realizing basic functions for implementing control and arbitration for general communication protocols.

Although each module may operate on a different clock frequency, all transfers occur synchronously within the clock environment of the StdChannel. Thus, StdChannel operation may be described as follows:

- Module A initiates a transfer using a synchronous send. Then, the StdChannel thread std_process_M_to_S transfers the Pdu from Msgbox A (attached to the StdChannel interface connected to module A) to Msgbox B (attached to the StdChannel interface connected to module B). Since send is synchronous, Module A is blocked until module B posts a receive command and a positive edge of the internal StdChannel clock occurs.

- Once the Pdu is transferred to Msgbox B, module B is able to receive the incoming Pdu (control signals and data) by posting a receive. Module B also synchronizes with its port by issuing a reply.

- Soon afterwards module B prepares its own Pdu (new set of control signals and data) and initiates a transfer towards module A using a synchronous send. Then, the StdChannel thread std_process_S_to_M transfers the Pdu from Msgbox B to Msgbox A. Module B remains blocked until module A posts a receive command and a positive edge of the internal StdChannel clock is reached.

- Once the Pdu is transferred to Msgbox A, module A is able to receive the incoming Pdu (control signals and data) by posting a receive. Module A also synchronizes with its port by issuing a reply and the protocol completes.

Assuming no module delays, e.g. due to delayed send, receive, or reply operation, clock synchronization, arbitration, or congestion, a complete StdChannel transaction requires 1 clock cycle in each direction, i.e. two clock cycles for completion of all StdChannel protocol communications.

5.3.4.4 Multi-Point StdBus Channel

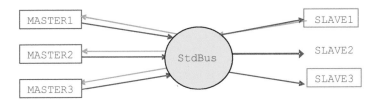

FIGURE 5.9: Multi-point, inter-module communication with StdBus

As shown in Figure 5.9, Standard Bus interface (described in `StdBus.h`) implements multi-point, bi-directional communication among several Master and Slave modules by forwarding channel-specific signals and user-defined data (size and type) among successive pairs of Master/Slave modules. For `StdBus`, the Master and Slave signals are not user-defined. While the pre-defined class `StdBusMasterCtrl` is used for connecting any Master module to the `StdBus`, no control signals are required by the Slave module, i.e. the corresponding `StdBusSlaveCtrl` class is empty, and thus normally only user-defined data can be transferred from Slave to Master; in general, control signal can be added through an inherited class which adds signals on top of the compulsory ones.

```
class StdBusMasterCtrl {
public:
N_uint8 priority;
N_uint address;
N_uint8 opcode;
N_uint be;
};
```

A Master module always initiates a `StdBus` transaction by transmitting (via a synchronous `send`) control signals and data to a particular Slave. Until the corresponding Slave module responds to the selected Master, and the Master acknowledges, no other Master may initiate another `StdBus` transaction to any Slave. Thus, `StdBus` channel implements many-to-many communication by allowing consecutive locking of channel resources by various one-to-one Master/Slave pairs. Each Master/Slave pair operates based on a simple, high-level, bi-directional, point-to-point inter-module communication protocol using two `StdBus` port-internal message boxes, identified as `MasterMsgbox` and `SlaveMsgBox`. These ensure the required synchronization between the bus process and the Master and Slave processes.

Although all modules operate on their own clock, `StdBus` transactions occur synchronously in the `StdBus` thread within the clock environment of the channel. `StdBus` operation and arbitration principles are described as follows.

- Each Master module may initiate a synchronous send transaction request to a specific Slave module by appropriately defining and initializing the corresponding user-defined data field, and the following `StdBus`-specific control fields in the Master Pdu:

    - opcode, as either `OCCN_write` or `OCCN_read`,
    - address mapped to the address space of a given Slave; the address space for each Slave is defined at construction time using the port function `set_slave_address_range` (see the example provided later),
    - packet priority, and

– byte enable defining which bytes within a packet are significant.

- **StdBus** arbitration, implemented as a thread within the **StdBus** channel process, selects a Master with a pending **OCCN_write** or **OCCN_read** request in a non-preemptive way. Selection is based on Master Pdu priority, and for equal priority, it is based on a simple round-robin order. At this point, **StdBus** is locked and all other Master requests become blocked until the selected communication transaction completes.

- After **StdBus** channel arbitration, the selected Master request becomes blocked, and its transaction fields (control signals and data) are saved to **MasterMsgbox** (attached to the **StdChannel** interface connected to Master modules).

- At the positive edge of the clock that the corresponding Slave is able to receive, information is transferred from **MasterMsgbox** to **SlaveMsgBox** (attached to the **StdBus** interface connected to Slave modules). After this operation, the **StdBus** thread is blocked until the Slave module posts a receive, in order to obtain the Master Pdu.

- Once the Slave module is able to obtain the Master Pdu using receive, it also synchronizes with the port by posting a reply.

- If the opcode is **OCCN_write**, then the corresponding Master module is ready to become unblocked when a positive edge of the internal **StdBus** clock occurs.

- If the opcode is **OCCN_read**, then additionally the following communication pattern is realized.

  – At first, after sending the acknowledgment, the Slave module prepares and transmits a Slave Pdu containing no control signals but only user-defined data (size and type) using **asend**. Notice that the Slave Pdu may contain different data, i.e. type and size, than the Master Pdu (either could be empty). Since asynchronous communication is used, the Slave module becomes unblocked when the channel, i.e. through **SlaveMsgBox**, obtains the data.

  – Then, the **StdBus** channel process transfers information from the **SlaveMsgBox** to the **MasterMsgBox**.

  – Finally, the Master module is able to obtain the Slave Pdu from **MasterMsgBox** using receive. It also synchronizes with its **StdBus** port by posting a reply operation.

In Figure 5.9, we connect three Master and three Slave modules through **StdBus**. For Slave modules, the function **set_target_address_range** provides the starting and end addresses that refer to the mapping of global addresses in the **StdBus** domain to a specific Slave module. Notice that no address range overlapping is allowed. Thus, a target may access a Slave module only if the address belongs to this address range. The channel process is able to access user-defined mapping from an address to a Slave by calling `get_slave_id_according_address(addr)`.

5.3.4.5   The `StdRouter`

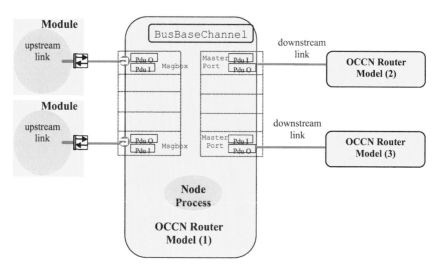

FIGURE 5.10: `StdRouter` internal structure

Starting from OCCN version 2.0, the OCCN library integrated a standard network model whose basic building block is the standard router model `StdRouter` shown in Figure 5.10. Next, we provide a short description of the main features of `StdRouter`; for a more complete description of `StdRouter` refer to Section 5.6 of the PhD thesis [271]. The basic modeling style of `StdRouter` reflects the one used for `StdChannel` and `StdBus`.

A standard router contains a parametric number of interfaces which are instantiated dynamically at binding time, and an array of master ports whose number depends on the router arity. The `BusBaseChannel` is the router base class. Each time a master port is bound to the standard router, function `get_master_if_pointer()` of the `BusBaseChannel` is called to obtain a new MasterIf interface pointer. This pointer stored in the master port enables primitive transfers, such as `send` or `asend`.

After network binding is completed, each router has instantiated a number of interfaces equal to its inbound links and a number of `MasterPort` objects equal to its outbound links. As reported in the router class declaration, the router micro architecture contains input and output buffers which represent the two router pipeline stages. These buffers are OCCN objects (CQueueObject) available in the OCCN library. Simple routing tables complete the router micro architecture. Input and output FSMs have been modeled using sequential C++ constructs.

The router node process is not a SystemC thread or method, but essentially it consists of sequential code based on three functions (input_th(), input_to_output_th(), and output_th()). These input/output switching functions model finite state machines of the standard router, performing arbitration and switch management by calling MsgBox internal functions, such as writing_completed() or authorize_reading(), that implement a message passing interface (see Section 5.3.4); refer to the router code for details.

More specifically, the input stage represents the passive stage of the model, since an array of (input) MsgBoxes whose arity is user-defined wait for commands from NI or router links (a NI link is considered identical to a router link). The output stage represents the active element whose purpose is to send flits to other routers and NI components and manage flow control according to the link protocol. Unlike StdBus and StdChannel, the output stage is implemented as an array of MasterPort interfaces. Similar to StdChannel, MasterPort signals used for link handshaking are not user-defined. Instead, a simple link protocol with signals modeled as plain variables has been deployed to demonstrate OCCN capabilities. Signals are described in struct cntrl_flit next (defined in std_network_data_type.h).

```
struct {
bool req;
bool ack;
bool eop;
bool sop;
} cntrl_flit;
```

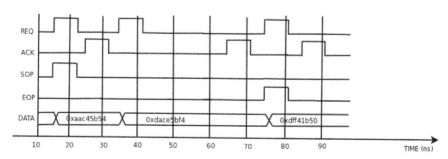

FIGURE 5.11: StdRouter link-level protocol

As shown in Figure 5.11, the link protocol relies on a simple request and acknowledgment mechanism. Aditional information is provided by the start of packet (`sop`) and end of packet (`eop`) signals. Data are simply modeled through a template of the Pdu data structure; this implicitly fixes the data path size.

Different node process functions corresponding to the entire NoC are scheduled by a global SystemC thread. Two important points are worth mentioning.

- Any topology can be easily built using the `StdRouter` provided by the OCCN library. The OCCN library in the supplementary CD contains network models for ring, 2-d mesh, and Spidergon STNoC topologies for demonstration purposes.

- By using static scheduling of input, switching and output stages and avoiding SystemC signals, the OCCN-based network can achieve very high simulation speed, while keeping clock accuracy [271].

The modeling techniques exploited with the more complex Spidergon STNoC router are built on top of the basic `StdRouter` modeling style. Spidergon STNoC models have been used in a number of ST Microelectronics platforms, providing fast and accurate network benchmarking; they particularly helped study and adapt the network architecture according to traffic requirements and services in line with ST Microelectronics division targets long before RTL component availability, proving OCCN effectiveness and fitness for purpose.

5.3.5   Case Study: OCCN Communication Refinement

Focusing on the user point of view, a case study for OCCN inter-module communication refinement and performance modeling based on a user-defined channel and the configurable ST Microelectronics' `STBus` is provided next.

This example shows how to develop the Adaptation Layer on top of the basic OCCN Communication API consisting of the `MasterPort` and `SlavePort` classes (see Figure 5.2). Using layered OCCN communication architecture, with each layer performing a well-specified task within the overall protocol stack, we describe a simplified transport layer, inter-module transfer application from a transmitter to a specified receiver. The buffer that has to be sent is split into frames. Thus, the TransportLayer API covers the OSI stack model up to the transport layer, since it includes segmentation. This API consists of the following basic functions.

- `void TransportLayer_send(uint addr, BufferType& buffer);` Notice that destination address `addr` identifies the target receiver and the buffer `buffer` to be sent. The `BufferType` is defined in the `inout_pdu.h` code block using the OCCN Pdu object.

- `BufferType* TransportLayer_receive();` This function returns the received buffer data.

We assume that the channel is unreliable. NoC is becoming sensitive to noise due to technology scaling towards deep submicron dimensions [35]. Thus, a simple stop-and-wait data link protocol with negative acknowledgments, called Positive Acknowledgment with Retransmission (PAR) (or Automatic Repeat Request, ARQ), is implemented [152] [351]. Using this protocol, each frame transmitted by the sender (I-frame) is acknowledged by the receiver using a separate acknowledgment frame (ACK-frame). A timeout period determines when the sender has to retransmit a frame not yet acknowledged.

The I-frame contains a data section (called payload) and a header with various fields:

- a sequence number identifying an order for each frame sequence; this information is used to deal with frame duplication due to retransmission, or reordering out-of-order messages due to optimized, probabilistic or hot-potato routing; in the latter case, messages select different routes towards their destination [129],

- a destination address field related to routing issues at Network Layer,

- an EDC (Error Detection Code) enabling error-checking at the Data Link Layer for reliable transmission over an unreliable channel [380], and

- a source_id identifying the transmitter for routing back an acknowledgment.

The ACK-frame sent by the receiver consists of only a header, with the following fields:

- a positive or negative ack field that acknowledges packet reception according to the adopted Data Link protocol, and

- a source_id identifying the receiver where to route the acknowledgment.

A frame is retransmitted only if the receiver informs the sender that this frame is corrupted, either through a special error code or by not sending an acknowledgment.

From the transmitter (Tx) side, the PAR protocol works as follows.

- Tx.1: send an I-frame with a proper identifier in the sequence field.

- Tx.2: wait for acknowledgment from the receiver until a timeout expires.

- Tx.3: if the proper acknowledgment frame is received, then send the next I-frame, otherwise re-send the same I-frame.

From the receiver (Rx) point of view, the PAR protocol works as follows.

- Rx.1: wait for a new I-frame from the Tx.

- **Rx.2**: detect corrupted frames using the sequence number (or possibly EDC).

- **Rx.3**: send a positive or negative **ACK-frame** based on the outcome of step **Rx.2**.

For simplifying our case study, we assume that no data corruption or packet loss occurs during the data exchanges. EDC is provided for a future, more complex implementation.

An OCCN implementation of the file transfer protocol can be based on inter-module communication between two SystemC modules (Transmitter and Receiver) through a synchronous, point-to-point OCCN channel, which is called **StdChannel**. This channel implements the timeout capability (see Section 5.3.4) and random packet loss by emulating channel noise.

5.3.5.1 OCCN Communication Refinement using **StdChannel**

The **StdChannel** is accessed through the OCCN Communication API, while the Transmitter and Receiver modules implement the higher-level Application API based on Adaptation Layer classes **MasterFrame** and **SlaveFrame** derived from **MasterPort** and **SlavePort**, respectively (see Figure 5.12). This SystemC-compliant approach allows design of the communication-oriented part of the application on top of the OCCN Communication API.

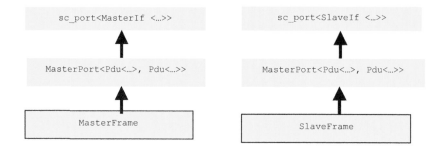

FIGURE 5.12: Inheritance of classes and specialized ports

Due to space limitation only the code for transmitter is provided below. Comments are introduced within each code block to illustrate important design issues.

With respect to data type definition, we define the buffer and frame data structures using the OCCN Pdu object. The buffer is an OCCN Pdu without header and a body of **Buffer_Size** number of characters. Mapping of the frame to an OCCN Pdu is shown in Figure 5.13. Assuming in-order transmission, the sequence number can be represented by a single bit. However,

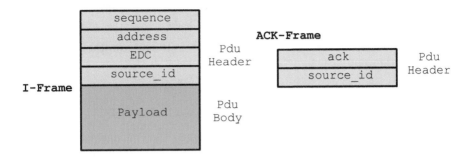

FIGURE 5.13: I- and `ACK-frame` data structures and Pdu objects

we assign a progressive, arithmetic sequence to each frame, partly for clarity reasons and for the possibility to support unordered transmissions in the future. Since `StdChannel` is point-to-point, addr is actually useless, but could be used in the future in a more general implementation. Moreover, a reserved number (`Error_Code`) in ack indicates an error code to distinguish, among all non-acknowledgment conditions, the ones relating to data corruption detected by EDC.

To simplify the code, we can assume that the data size of the `StdChannel` equals the frame size, i.e. Pdu sent by the OCCN Communication API is directly the `I-frame` in the transmitter to receiver direction and the `ACK-frame` in the opposite.

The `inout_pdu.h` code block provides the Pdu type definitions for inter-module communication.

The Transmitter module implements the `sc_thread action_tx`. A buffer filled with random letters in the range 'A' to 'Z' is sent through the channel by calling the application API `TransportLayer_send` function through the `MasterFrame sap_tx` access port. This operation is repeated `Nb_Sequences` times. The Transmitter module interface includes:

- the `MasterFrame` definition, which is the same as in the Receiver module; thus it is obtained directly from `inout_pdu.h`,

- the transmission layer interface definitions, defined in `MasterFrame.h`,

- the thread name and action (`action_tx`) routine,

- the thread definition,

- the transmission layer name, as well as

- other internal objects and variables, such as buffer.

The Transmitter module provides a transmission layer interface described in code blocks `MasterFrame.h` and `MasterFrame.cc`. This layer defines a very simple communication API based on the `TransportLayer_send` function; its behavior can be summarized into two main actions:

- segmentation of the buffer into `I-frames`, with the relevant header construction and insertion; this action exploits Pdu class operators.

- sending the `I-frame` according to the PAR protocol.

Finally, the `main.cc` references to all modules. It essentially

- instantiates the simulator `Clock` and defines the timeout delay,

- instantiates the SystemC Receiver and Transmitter modules,

- instantiates the `StdChannel`,

- distributes the Clock to the Receiver and Transmitter modules,

- connects the Receiver and Transmitter modules using a point-to-point channel (and in more general situations a multi-point channel), and

- starts the simulator using the `sc_start` command.

5.3.5.2   OCCN Communication Refinement using STBus

This Section explains communication refinement if the proprietary STBus is used instead of the generic OCCN `StdChannel`. A TLM cycle-accurate model for STBus has been developed using OCCN. The model provides all OCCN benefits, such as in simplicity, speed, and protocol in-lining. System architects are currently using this model in order to define and validate new architectures, evaluate arbitration algorithms, and discover trade-offs in power consumption, area, clock speed, bus type, request/receive packet size, pipelining (asynchronous/synchronous scheduling, number of stages), FIFO sizes, arbitration schemes (priority, least recently used), latency, and aggregated throughput.

We next illustrate important concepts in communication refinement and design exploration using the STBus model. Similar refinement or design exploration may be applied to AMBA AHB or APB bus models. These models have also been developed using OCCN. For further information regarding STBus and AMBA bus OCCN models please refer to [80] [301] [302].

Refinement of the transport layer data transfer case-study is based on the simplest member of the STBus family. STBus Type 1 acts as an RG protocol, involves no pipelining, supports basic load/store operations, and is targeted at modules with low complexity, medium data rate system communication requirements.

Figure 5.14 shows the simple handshake interface of STBus Type 1 [301] [302]. This interface supports a limited set of operations based on a packet containing one or more cells at the interface. Each request cell contains various fields: operation type (`opc`), position of last cell in the operation (`eop`),

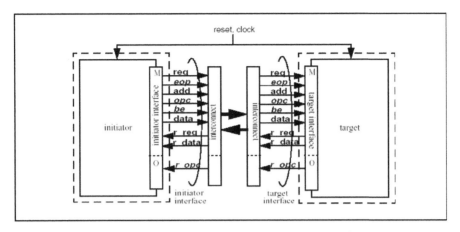

FIGURE 5.14: STBus Type 1 initiator and target interfaces

address of operation (`add`), data (`data`) for memory write, and relevant byte enable signals (`be`). The request cell is transmitted to the target, which acknowledges accepting the cell by asserting a handshake (`r_req`) and sending back a response packet. This response contains a number of cells, with each cell containing data (`r_data`) for read transactions, and optionally error information (`r_opc`) indicating that a specific operation is not supported, or that access to an address location within this device is not allowed. STBus uses r_opc information to diagnose various system errors.

The API functions `TransportLayer_send` and `TransportLayer_receive` do not change in terms of their semantics and interfaces, but the Adaptation Layer must implement the functionality required to map the Application API to the STBus intrinsic signals and protocol. Thus, within the Adaptation Layer, we provide an STBus-dedicated communication API based on the `MasterPort` and `SlavePort` API. Due to space limitation, the code for this refinement is not provided. However, we abstract some of the key ideas used in the process.

According to Figure 5.14, the frame must be mapped to an `STBus_request` Pdu describing the path from initiator (transmitter) to target (receiver), and an `STBus_response` Pdu representing the opposite path. In particular,

- the payload corresponding to STBus write data lines (`data`) is mapped to the body of the `STBus_request` Pdu,

- buffer segmentation and reassembly are implemented as in `StdChannel` by exploiting the `>>` and `<<` operators of the OCCN Pdu object (the payload size is the same),

- the destination address corresponds to the STBus addr signal (which is a part of the `STBus_request` Pdu header), and

- the extra bits for the EDC are implemented as extra lines of the write data path.

With Type 1 protocol, completion of the initiator send means that the target has received the data. Thus, we avoid explicit implementation of the sequence, source_id (both fields) and ack fields in Figure 5.14.

Furthermore, since STBus guarantees a transmission free of packet loss, normally no timeout features are required. However, deep submicron effects may make the STBus a noisy channel. Thus, we may extend the basic idea of a PAR protocol, i.e. EDC with retransmission of erroneous data, to a real on-chip bus scheme [36] [282]. In this case, the r_opc signal, represented as a header field of the STBus_response Pdu, may trigger an error code to the transmitter (initiator, in STBus terminology). This code may be used to determine if the same transaction must be repeated.

Since there is no direct access to the signal interface or the communication channel characteristics, we do not need to modify Transmitter or Receiver application modules, or the relevant test benches. Thus, we achieve module and tester design reuse at any level of abstraction without any rewriting and without ripping and rerouting communication blocks. This methodology facilitates OCCA design exploration through efficient hardware/software partitioning and testing the effect of various functions of bus architectures. In addition, OCCN extends state-of-the-art in communication refinement by presenting the user with a simple, flexible and compositional approach that enables rapid IP design and system-level reuse.

5.3.5.3   System-Level Performance

OCCN methodology for collecting statistics from system components can be applied to any modeling object. For advanced statistical data, which may include preprocessing, one may also directly use the public OCCN statistical classes. In order to generate basic statistics information appropriate enable_stat_ function calls must be made, usually from within the module constructor. For example, below we show the function call for obtaining write throughput (read throughput is similar) statistics.

```
// Enable statistics collection in [0,50000]
// with number of samples = 1
enable_stat_throughput_read(''statbox'', 0, 50000, 1,
''Simulation Time",''Average Throughput for Write Access");
```

Considering our transport layer data transfer case study without taking into account retransmissions, we measure the effective throughput in I-frame transfers in Mbytes/sec. Assuming packet-loss transmission, and a receiver that provides an acknowledgment after every clock cycle, i.e. a frame is transmitted every 2 cycles, the StdChannel can transmit a payload of 4 bytes during every (10ns) clock cycle. Thus, the maximum throughput is 200Mbytes/sec.

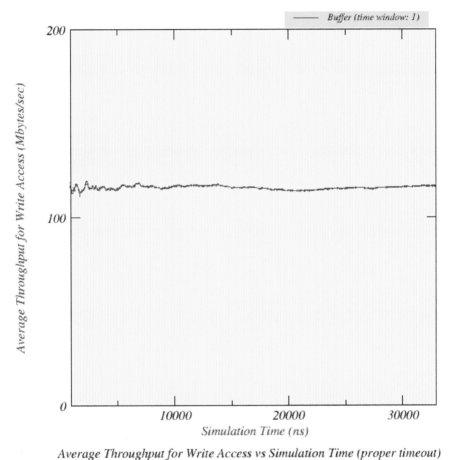

Average Throughput for Write Access vs Simulation Time (proper timeout)

FIGURE 5.15:   Performance of transport layer protocol (proper timeout)

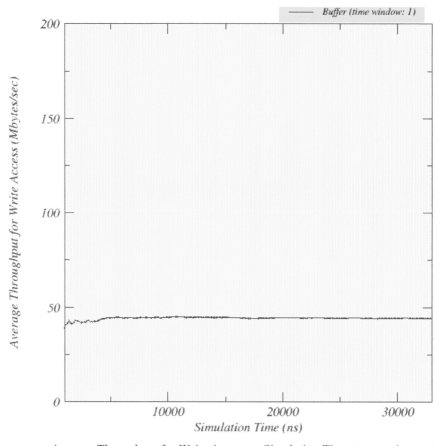

Average Throughput for Write Access vs Simulation Time (wrong timeout)

FIGURE 5.16:   Performance of transport layer protocol (wrong timeout)

Figures 5.15 and 5.16 assume a receiver with random response latency (not greater than 3 clock cycles) and an unreliable connection. In the first graph, an appropriate timeout is chosen according to the receiver latency, while in the second graph, the chosen timeout is too short, thus a high number of retransmissions occur. This obviously decreases the performance of the adopted ARQ protocol.

5.3.5.4 Design Space Exploration using OCCN

In order to evaluate the vast number of complex architectural and technological alternatives the architect is equipped with a highly-parameterized, user-friendly, and flexible OCCN methodology. This design exploration methodology is used to construct an initial architectural solution from system requirements, mapping inter-module communications through appropriate bus configuration parameters. Subsequently, this solution is refined through an iterative improvement strategy based on domain- or application-specific performance evaluation based on analytical models and simulation. The proposed solutions provide values for all system parameters, including configuration options for increasingly sophisticated multiprocessing, multithreading, prefetching, and cache hierarchy components.

Performance estimation is normally based on mathematical analysis models of system reliability, performance, and resource contention, as well as stochastic traffic models and actual benchmarks for common applications. Common embedded system benchmarks involve integer and fixed point computations, pipelining, and usually irregular communication patterns. Examples include commercial networking, multimedia, concurrent and/or hierarchical finite state machine problems, graph traversal (e.g. sorting), combinatorial logic (e.g. binary decision diagrams, encryption), and finite or infinite impulse response digital signal filtering. Performance evaluation must always take into account future traffic requirements arising from new applications, scaling of existing applications, or evolution of communication networks.

After extensive architecture exploration, ST Microelectronics exploits a reuse-oriented design methodology for implementing a generic interconnect subsystem. The system-level configuration parameters generated as a result of architectural exploration are loaded onto the Synopsys tools coreBuilder and coreConsultant. These tools integrate a preloaded library of configurable high-level (soft) IPs, such as the STBus interconnect; IP integration is performed only once using coreBuilder. CoreConsultant uses a user-friendly graphical interface to parameterize IPs, and automatically generate a gate-level netlist, or a safely configured and connected RTL view (with the correct components and parameters), together with the most appropriate synthesis strategy. Overall SoC design flow now proceeds normally with routing, placement, and optimization by interacting with various tools, such as Physical Compiler, Chip Architect, and PrimeTime.

Architecture exploration using OCCN allows the designer to rapidly assemble, synthesize, and verify a NoC through a methodology that uses pre-designed IP for each bus in the system. This approach dramatically reduces time-to-market, since it eliminates the need for long redesigns due to architecture optimization after RTL simulation.

5.3.5.5   Power Analysis Tools

In addition to high-level performance tools, there is a growing trend towards applying power estimation techniques for energy-efficient design. In fact, as we progress from high- to low-level design, power modeling, analysis and optimization tools become more accurate but also about an order of magnitude slower, thus being able to handle much smaller circuits. Typically transistor level tools produce results that are within 5-10% of measured results, gate-level tools are accurate to about 10-15% and architectural level tools (e.g. Verilog/VHDL RTL tools) are typically within 20-25% of silicon.

With feed-forward design methodology, design does not progress to the lower abstractions, i.e. synthesis or subsequent layout, until the architecture satisfies the power specification at the higher level. Thus, we can derive rapid and more accurate architecture-level power estimation models by annotating power data obtained from structural or behavioral RTL simulation into high-level models, such as cycle-accurate TLM.

These high-level models will often include analytical device macro models, e.g. instruction-based lookup tables for processors [358] [122] and custom circuits [276]. In addition, appropriate data representation, algorithm design and implementation are key considerations for achieving energy efficiency in Multicore SoC designs with computation, storage, and communication components. Notice that although while computation and storage energy greatly benefit from device scaling, the energy for global on-chip communication does not scale and requires increasingly higher energy consumption, especially as we move to many cores. Hence, communication energy minimization and roadmapping are of growing concern in future technology [186].

In this respect, OCCN is able to address design of emerging energy-aware Multicore SoCs through existing high-level SystemC-based power modeling approaches that are

- simple enough, i.e. they do not increase model development design time,

- flexible to allow dynamic manipulation of power parameters,

- fast enough to perform efficient simulation and design space exploration,

- accurate to allow the energy comparison of several architectures and the application of dynamic power management techniques,

- powerful enough to allow power modeling of all SoC components during simulation, and finally

- open source, due to GNU license requirements.

Existing high-level power modelling methodology includes numerous commercial products and Academic research tools. Unfortunately, most of these approaches are currently not open source.

Commercial high-level power estimation, analysis and optimization tools concentrate on power-driven synthesis and gate level power. These tools mainly come from start-up companies and SMEs. Some of these methods provide not only the total energy consumption at the end of the simulation, but also its evolution in time. Tools include Simunic [319], web-based Joule-Track [321], Avalanche [140], Lajolo [200], Powerchecker [53] and Orinoco [70] [332] and HyPE [220].The first four tools include power models for processor with an ISS (Instruction Set Simulator). Lajolo and Powerchecker use slower RTL simulation incorporating power estimation model for hardware component. Hardware-software co-simulation is performed by linking Avalanche and Lajolo. BullDast's Powerchecker works on a mixed RT-gate level description obtained through source HDL analysis, elaboration and hardware inferencing [277]. Design objects are annotated with real switching activities obtained through RTL simulation. Chipvision's Orinoco is a system-level design space exploration tool that estimates performance and power in running specific algorithms (specified in C/C++) on different architectures [332]. The algorithm compiles to a hierarchical control data flow graph (CDFG) which describes the expected circuit architecture without resorting to complete synthesis. CDFG nodes represent power-characterized operations, edges represent control and data dependencies among operations, and nested procedure calls correspond to transitions between successive hierarchy levels [6]. Compositional rules compute the total cost of a complex CDFG depending on implementation style. HyPE is a high-level simulation tool that uses analytical power macro-models for fast and accurate power estimation of programmable systems consisting of datapath and memory components [220].

Power-Kernel is an Academic object-oriented library for SystemC 2.0, which allows simple introduction of a power macro model in SystemC at RTL level of a complex design [79]. PK achieves much higher simulation speed than lower-level power analysis tools. High-level model instrumentation is based on a SystemC class that uses advanced dynamic monitoring and storage of I/O signal activity of SoC blocks with put_activity and get_activity functions [57]. Both constant power models and more accurate regression-based models with a linear dependence on clock frequency, gate and flip-flop switching activity are used. As an example, dynamic energy estimation of the AMBA AHB bus is decomposed into arbiter, decoder and multiplexing logic for read and write operations (master to/from slave). The latter operations are estimated to control over 84% of the total dynamic power consumption. Similar power instrumentation techniques for synthesizable SystemC code at RTL level are described in [389].

5.3.5.6   Conclusion and Future Extensions

Network on-chip is becoming a critical determinant of system-level metrics, such as system performance, reliability, and power consumption. The development of SystemC as a general-purpose programming language for system modeling and design enables the application of C++ object-oriented technology for modeling complex Multicore architectures, involving thousands of general-purpose or application-specific PEs, storage elements, embedded hardware, analog front-end, and peripheral devices. The OCCN project is aimed at developing new methodology and tools for the design of network on-chip for next generation high-performance Multicores.

NoC design flow calls for the development of new methodology and efficient algorithms for automatic design space exploration of high-performance network on-chip. Within this scope, the focus is on system-level performance characterization and power estimation using statistical (correlated) macros for NoC models, considering topology, data flow, and communication protocol design. These models would eventually be hooked to the OCCN framework for SystemC simulation data analysis and visualization. Within this scope, power estimation macros have been used within OCCN, mostly within an AHB power estimation framework [57]. This means that a cycle-accurate OCCN AHB bus model integrated high-level power estimation functions derived from VHDL power tools, such as those described in the previous Section.

While OCCN focuses on providing important modeling components and appropriate methodology, sophisticated tools are necessary for efficient NoC design space exploration. For example, interactive and off-line visualization techniques including advanced monitoring features, such as generation, processing and presentation, would enable improved performance modeling and analysis. In addition, platform performance metrics can be especially helpful in monitoring and improving simulation efficiency, e.g. through data partitioning or dynamic load balancing.

Chapter 6

Conclusions and Future Work

This book has addressed emergence of the Multicore era from the challenging perspective of on-chip communication.

The Multicore concept has been defined and analyzed from different perspectives, considering application and market requirements and focusing on software and hardware architecture, while also discussing technological issues.

Current homogeneous Multicore processors consist of several identical cores on the same die, interconnected through a hierarchy of cache and local memory, and sharing an external memory through one or more memory controllers. These architectures are open and generic to execute different applications not known a priori.

Alternative Multicore System-on-Chip architectures are customized to embedded consumer market applications. In these systems several heterogeneous cores, with a generic or specialized instruction set, are coupled with hardware accelerators running at lower speed in order to offer the right level of performance, power dissipation, design flexibility, and cost.

Nowadays, advanced SoC in the embedded domain and processors in the PC market are implementing different flavors of the Multicore concept. In the future, an evolution towards convergence is foreseen, with an open, configurable platform based on a combination of heterogeneous cores.

During this shift from computation- to communication-centric platforms, at the heart of these architectures, on-chip communication is required to evolve in order to match increased application performance, power consumption, protocol, and manufacturability requirements.

Motivation along with a detailed description of historical advances in this domain, from on-chip buses and cascaded crossbars to recent Network-on-Chip, are provided. Moreover, as stated in this book, a further revolutionary shift will occur soon towards advanced on-chip communication components that we defined as Interconnect Processing Units (IPUs). In our vision, the IPU is an on-chip communication network that includes hardware and software components, which jointly implement Multicore programming language primitives and provide low-level platform services to enable advanced features in modern applications of feature-rich devices. The IPU is a communication component which is software programmable at runtime, hardware configurable and with extensible services at design time which can be customized depending on specific requirements. IPUs will play a crucial role in future configurable and communication-centric, heterogeneous Multicore architectures, by combining host processors, such as ARM, configurable cores, such as ARC and Tensilica, and IPUs, such as Spidergon STNoC.

Driven by this vision, the main architectural issues surrounding design of the Spidergon STNoC IPU have been addressed. As an industrial program carried out at the Advanced System Technology group of ST Microelectronics, the Spidergon STNoC on-chip interconnect technology is a first attempt at designing an IPU for an advanced Multicore platform. At a first glance, the Spidergon STNoC IPU is a flexible on-chip communication platform based on a pseudo-regular network infrastructure implementing a set of communication primitives and programmable services for different of applications.

We have described the cornerstones of the network architecture, namely topology, switching strategy, communication protocol layering, packet structure, routing algorithm, buffer management and flow control, QoS support, hardware building blocks and services. Our architecture and design strategy is motivated and especially applied to hardware components and basic network choices in order to illustrate how it is possible to move in a pragmatic and concrete way towards the IPU concept.

As an aggressive and innovative example of this architecture philosophy, Spidergon STNoC leverages on regularity of Spidergon topology, while featuring practical topological customizability. Hence, Spidergon topology is a cost-efficient solution for the embedded SoC context, between a simple and low performance ring and a highly connected mesh. Although specific embedded applications normally require customized communication infrastructures, heterogeneity can be captured by aggregation and composition of network graphs in limited hierarchies. In fact, Spidergon STNoC fills the gap between regularity and heterogeneity through three main topological features. First, the regular topology can be customized and simplified depending on application traffic requirements. Second, the injection and extraction traffic can be aggregated in a flexible way. Finally, multiple hierarchical regular networks can be connected to each other to reflect a functional decomposition of on-chip communications.

Unlike existing NoC architectures which are either topology-independent, i.e. supporting any possibly irregular structure, or based on a fixed regular graph, Spidergon STNoC fills the gap between these two approaches, trading off regularity and customizability.

Design of the low-cost Spidergon STNoC architecture follows simple, clean and well-defined communication layering based on three hardware components: a physical communication link, a wormhole router, and a network interface to plug cores, hardware IPs and memories. The fourth module of the Spidergon STNoC platform is the software component which implements IPU services. Although this component is currently available, it is currently further being refined and will be described in detail in future publications.

In order to address all Multicore application requirements, Spidergon STNoC technology provides a set of programmable services that implement the basis for feature-rich devices. These services may be instantiated or not depending on the real target application, resulting in an extensible on-chip communication infrastructure. Among the services currently provided, the main ones are security, power management, advanced communication primitives, QoS, and path redundancy.

The hardware/software communication platform within the IPU is associated to a specific use methodology. In this book, we have focused on our open-source OCCN simulation framework, which refers to system-level NoC modeling, design space exploration, and architecture refinement.

Driven by our vision for migration from on-chip communication architectures towards IPUs, the Spidergon STNoC program will evolve in several directions. Hereafter, we briefly outline the main issues which are currently explored and will be completed in the future.

- Enriching the IPU programmability portfolio, both in terms of supported communication primitives and low-level platform services.

- Developing and testing advanced physical design components and methodologies on silicon.

- Completing the EDA toolkit for IPU design space exploration, targeting especially for mapping, configuration, and simulation.

- Defining and exploiting new design and verification methodologies particularly targeted to highly software programmable, hardware configurable and extensible IPU blocks.

6.1 Enhanced IPU Programmability Portfolio

New IPU communication primitives and low-level platform services will be introduced in the future on top of the Spidergon STNoC on-chip network infrastructure.

In terms of communication primitives, we plan to support sophisticated semantics for shared memory, message passing, and streaming contexts. For example, we plan to provide native protocol support for cache coherency to enable relatively small, Spidergon STNoC clusters of processors or specialized cores to be integrated consistently as sub-networks in the overall on-chip communication infrastructure.

We will also focus on mapping low latency synchronization functions to improve efficiency of inter-processor communication. In particular, as already stated, we study support of an ultra-lightweight version of Active Messages, to associate a small amount of computation through an appropriate handler upon reception of a network transaction. In this simple remote invocation, the handler type is carried by the network transaction itself to be executed immediately upon arrival, thus eliminating overhead from traditional interrupt interpretation schemes.

Moreover, in the context of shared memory protocols, OCP 3.0 (forecast for year 2008) will be supported at the Network Interface IP socket. All aspects of efficient interoperability among different protocols will be further investigated and completed.

It is likely that future network interface implementations will be hybrid schemes providing low latency for delay-bound applications, while including low-cost and reasonably efficient buffering as a fallback mode, as in user-defined messages in FUGU [221] [69] [244]. In this case, bulk message transfers can be handled by a direct memory access (DMA) mechanism supported by a small, message queue and a simple DMA engine.

In the near future, Spidergon STNoC will increase the number of low-level platform services integrated on top of its infrastructure, especially for ensuring efficient management of hardware resources depending on application.

We believe that it is mandatory to leverage on the physical distribution of the interconnect over the chip and exploit this at system-level. For instance, on-silicon system performance tracking and debug could help in testing complex silicon chips, especially during first product generations. In addition, the capability to monitor the system state, e.g. traffic load, latency, core power state, security attacks, and system error conditions in all chip regions can enable efficient centralized or distributed resource management, as described below.

Unlike programmability of SoC which is currently possible only through the processor instruction set, the IPU intends to offer RTOS and middleware the capability to change silicon functionality through a set of programmable low-level platform services of the on-chip interconnect. Spidergon STNoC will continue evolving in this direction, e.g. by increasing the number of features exposed to software and simultaneously increasing the complexity of the IPU software component to enable automatic (or semi-automatic) platform management.

In Spidergon STNoC, these services are controlled at the boundary of the on-chip network infrastructure, allowing for simple and efficient software control. We plan to extend hardware support of new services in the main building blocks, especially in the router and network interface. We also plan to focus on the development of light software components that can implement a part of service management.

By monitoring hardware platform states and receiving inputs from application platform states, the software component integrated in the Spidergon STNoC IPU may reprogram certain specific low-level platform services. For instance, depending on network monitoring of real traffic conditions in application use cases, software components could modify the routing scheme, QoS, or power management to redistribute the load, decrease the latency, or minimize the total chip energy consumption.

Programmability of the Spidergon STNoC IPU could evolve into another exciting direction. Network topologies can eventually become highly polymorphic, i.e. their configuration and properties can be radically changed during runtime by simply allocating extra programmable routers connected together through additional direct data paths to form a redundant lattice. We are investigating how programmable routers can be configured during runtime through special switch control functions. Polymorphic network topologies can be algorithmically specialized by exploiting on-the-fly mapping of the lattice to application tasks with different sizes and characteristics, e.g. through tree or mesh-like patterns which can be allocated either spatially or temporally [324]. Implementing such methods would require a shift towards innovative high-level programming models, supported by intelligent model-driven tools.

6.2 IPU Physical Design

The Spidergon STNoC has been designed to allow better manufacturability in future nanometer technologies. We currently approach the design of silicon demonstrators of the Spidergon STNoC technology and expect important results to become available in the near future.

As an exciting challenge, the IPU is a unique component from the software platform perspective, while from the physical implementation point of view it must address CMOS technology scaling by adopting a highly distributed structure. In order to exploit best its physical nature as a pseudo-regular network, novel methods of SoC physical integration have to be defined and investigated, targeting a quick assembly of IP blocks around the IPU in plug-and-play fashion, thus saving tremendous efforts in reaching chip time closure. We are currently investigating these new methods in several physical design trials, considering the recent CMOS technologies of 45nm and 32nm.

Part of this manufacturing-aware approach is the capability to mix different physical links in the same Spidergon STNoC instance. Evolution in the definition of new advanced links is progressing in several directions. To cope with synchronization, relaxing clock distribution constraints and considering clock variability is one of the main concern we are currently facing. We will evolve links into sophisticated schemes that ensure an asynchronous communication through a reduced set of wires, in order to implement in-package communications or sub-network bridging.

We will evolve sophisticated links that ensure asynchronous communication through a reduced set of wires. These protocols will be based on FPGA prototyping scenarios, inter-die in-package communications and sub-network bridging.

Moreover, the physical link will integrate fault tolerance schemes, essentially based on optimized error detection and correction codes, to increase reliability of the transmission at least for some strategic parts of the transmitted packet information. These techniques at link level will be combined with end-to-end mechanisms managed at the Network Interface to notify and eventually recover error conditions in data transfers.

6.3 IPU Design Tools

The IPU is a complex component which not only can be programmed at runtime, but it can also be configured and extended at design time. In order to achieve this property on Spidergon STNoC, we have mainly worked on creating a flexible, easily customizable hardware infrastructure.

At the same time, it is mandatory to rely on tools for helping the user customize the IPU architecture template. As far as tools are concerned, we have dedicated a great effort in modeling methodology and simulation of on-chip communication networks by developing the open-source OCCN framework.

We are currently focusing on higher level tools that help in design space exploration, as well as in the definition and generation of alternative instances of Spidergon STNoC. The innovative approach that we follow leverages on static analytical models, instead of traditional dynamic simulation at different levels of abstraction. In this respect, we are working towards developing a comprehensive graphical environment in which the IPU user (i.e. the Multicore SoC architect) can preliminary evaluate topology, routing, and service alternatives through analytical functions. This framework obviously helps in tuning the best architectural configuration in terms of system parameters, such as flit size, queue size, type of physical links, as well as for generating a simulation model for validation or synthesis.

To evaluate alternative architectures and define the optimal one, especially in the context of the Spidergon STNoC pseudo-regular structure, it is not necessary to rely on very accurate latency and bandwidth values or post-synthesis area and power consumption metrics. Instead, simpler analytical models that are fast to compute can drive most of the strategic choices in the early stages of architecture definition. Thus, in fact, latency, bandwidth, and other performance values, as well as area and power metrics, can be computed at different levels of the design flow.

Automated mapping of application traffic onto Multicore system resources is based on application modeling and graph embedding onto a NoC topology model which abstracts the actual NoC architecture onto which the application is deployed and run. Mapping is influenced by static or dynamic requirements, such as amount of memory for all tasks assigned to the same node, application latency, bandwidth, energy metrics, and termination deadlines for hard real-time tasks. It has been estimated that SoC performance varies by up to 250% depending on NoC design, and up to 600% depending on communication traffic [199], while NoC power dissipation is also reduced by more than 60% by using appropriate mapping algorithms [154].

Previous research has addressed energy-efficient static embedding of different applications (e.g. multimedia or DSP) on conventional NoC topologies, such as 2-d mesh, torus, or heterogeneous NoC architectures [296] [383] [392] [154], as well as related embedding tools, such as SUNMAP [246] [247]. For a particular Multicore application, these tools not only address selection of the NoC topology, routing algorithm, partitioning heuristic, or dynamic voltage/frequency level for all links [314], but they also serve as basis for developing appropriate operating system support. Other compiler tools target synthesizable, cycle- and bit-accurate, SystemC code generation for the optimal NoC topology, with generic routers and network interfaces configured with the best routing strategy [168].

Despite static analysis and automated mapping, we believe that cycle-accurate simulation is necessary to complete the architecture definition, verify the defined instance, and eventually refine it in terms of key design parameters based on performance, power and relative cost characteristics; notice that refinement may take into account accurate IP traffic generators and sophisticated memory controller models.

High-level power estimation based on component models with power state abstractions, analytical equations or lookup tables is at its infancy, since application test benches are unavailable or only partially known during the early design phases. Nevertheless, high-level power estimation has been gaining momentum in the industry, as the new entry point in the design flow. It is orders of magnitude faster than RTL simulation, has a much shorter modeling time, and yet it is sufficiently (near RTL) accurate to achieve considerable power savings during design space exploration.

In this respect, high-level bit- and cycle-accurate transaction- or behavioral-level NoC modeling (e.g. in SystemC2.0+) can be used for evaluating performance versus power consumption tradeoffs, as well as critical architectural and algorithmic characteristics, such as power saving, pipelining, and parallelization. Total dynamic power consumption can be estimated by multiplying the number of transactions and/or bit transitions for each (memory, register, interface, control and data path) component with appropriate roadmap-based basic bit energy coefficients; e.g. power per transaction, operation, or gate transition. Transactions are abstract operations applied to a component, while bit transitions reflect individual bit changes from past to current data payloads that take place as a consequence of these operations. Our research will eventually address the need to increase the abstraction level of these power models by associating an energy cost to each IPU communication primitive or low-level platform service.

6.4   IPU Design and Verification Methodology

Apart from the methodology and tools for IPU design exposed to the Spidergon STNoC user which have been discussed in Section 6.3, IPU design requires novel methods to support efficiently development and verification of this software programmable, hardware configurable and extensible block.

Although modification of different types of architectural parameters, such as flit size or queue size, can be managed efficiently by traditional HDL languages, the concept of optionally adding primitives and service support as hardware or software extensions requires great flexibility on the adopted description language. The goal is to guarantee the best optimized performance in terms of area and operating frequency, independent of the selected IPU configuration and extension. With respect to methods currently available in the EDA market or by companies specializing in configurable cores, the unique challenge that we are facing is the double degree of configurability provided by the IPU. Indeed the IPU is built as a topological composition of several small elementary building blocks. Since configurability not only is applied to each building block, but also to the entire IPU structure, i.e. the topology is configurable, simple control of overall IPU flexibility as a unique object is challenging.

References

[1] Abdennadher, A. and Feng, T.Y., "On rearrangeability of Omega-Omega networks", *in Proc. Int. IEEE Conf. Parallel Proc.*, Vol. 1, 1992, pp. 159–165.

[2] Abolhassan, F., Keller, J., and Paul, W.J., "On the cost-effectiveness of PRAMs", *Acta Informatica*, **36 (6)**, 1999, pp. 463–487.

[3] Adve, S.V. and Gharachorloo, K. "Shared memory consistency models: a tutorial", *IEEE Trans. Computers*, **C-45 (12)**, 1996, pp. 1145–1155.

[4] Agarwal, A., Bianchini, R., Chaiken, D., et al., "The MIT Alewife machine: architecture and performance", *in Proc. IEEE Symp. Comp. Arch.*, 1995, pp. 2–13.

[5] Agarwal, A., Bianchini, R., Chaiken, D., et al., "The MIT Alewife machine", *in Proc. IEEE*, 1999.

[6] Aho, A., Sethi, R., and Ullman, J., *Compilers: Principles, Techniques and Tools*, Addison-Wesley, 1988.

[7] Albonesi, D.H. and Koren, I., "STATS: a framework for microprocessor and system-level design space exploration", *J. Syst. Arch.*, **45**, 1999, pp. 1097–1110.

[8] Allen, R. and Garlan, D., "A formal basis for architectural connection", *ACM Trans. Soft. Engin. and Methodology*, **6 (3)**, 1997, pp. 213–249.

[9] Altera, "Avalon bus specification: reference manual", 2003. Available at http://www.altera.com

[10] ARM Amba Bus, see http://www.arm.com

[11] Ambric, see http://www.ambric.com

[12] AMD, see http://www.amd.com

[13] Anderson, T., Owicki, S., Saxe, J.B., and Thacker, C., "High speed switch scheduling for local area networks", *ACM Trans. Comput. Syst.*, **11 (4)**, 1993, pp. 319–352.

[14] Andriahantenaina, A. and Greiner, A., "Micro-network for SoC: implementation of a 32-port SPIN network", *in Proc. Design Automation and Test in Europe Conf.*, 2003, pp. 1128–1229.

[15] Andriahantenaina, A., Charlery, H., Greiner, A., et al. "SPIN: a scalable, packet switched, on-chip micro-network", *in Proc. Design Automation and Test in Europe Conf.*, 2003, pp. 70–73.

[16] Android, see http://code.google.com/android/

[17] ARM Amba Bus, see http://www.arm.com

[18] ARC, see http://www.arc.com

[19] Arora, S., Leighton, F.T., and Maggs, B.M., "On-line algorithms for path selection in a nonblocking network", *SIAM J. Comput.*, **25 (3)**, 1996, pp. 600-625.

[20] Asanovic, K., Bodik, R., Catanzaro, B.C., et al., "The landscape of parallel computing research: a view from Berkeley", *Techn. Rep. UCB:EECS-2006-183*, Dept. Electr. Engin. Comp. Sci., University of California, Berkeley, 2006.

[21] Ashenden, P.J., Mermet, J.P., and Seepold, R. (eds), *System On Chip Methodology and Design Languages*, Kluwer Academic Publisher, 2001.

[22] Bassalygo, L.A. and Pinsker, M.S., "Complexity of an optimum nonblocking switching network without reconnections", *Problems of Inform. Transm.*, **9**, 1974, pp. 64–66.

[23] Bacon, D.F., Strom, R.E., and Tarafdar, A., "Guava: a dialect of Java without data races", *in Conf. Obj. Oriented Progr., Syst., Lang., and Appl. (ACM SIGPLAN)*, **35**, 2000, pp. 382–400.

[24] Balfour, J.D. and Dally, W.J., "Design tradeoffs for tiled CMP on-chip networks", *Int. Conf. Supercomputing (ICS)*, 2006, pp. 187–198.

[25] Bainbridge, W.J., Toms, W.B., Edwards, D.A., and Furber, S.B., "Delay-insensitive, point-to-point interconnect using m-of-n codes", *in Proc. Int. IEEE Symp. on Asynchr. Circ. and Syst.*, 2003.

[26] Bainbridge, W., "Asynchronous system-on-chip interconnect", *Ph.D. Thesis*, Computer Science, University of Manchester, England, 2000.

[27] Bainbridge, J. and Furber, S.B., "Chain: a delay-insensitive chip area interconnect", *IEEE Micro*, **22 (5)**, 2002, pp. 16–23.

[28] Balarin, F., Chiodo, N., Giusto, P., et al., *Hardware-Software Co-design of Embedded Systems: The Polis Approach*, Kluwer Academic Publishers, 1997.

[29] Baran, P., "On distributed communication networks", *IEEE Trans. Comm.*, **C-12 (1)**, 1964, pp. 1–9.

[30] Beneš, V.E., *Mathematical Theory of Connecting Networks and Telephone Traffic*, Academic Press, 1965.

[31] Ben-David, S., Borodin, A., Karp, R., Tardos, G., et al., "On the power of randomization in on-line algorithms", *Algorithmica*, **11 (20)**, 1994, pp. 2–14.

[32] Benini, L., Bertozzi, D., Bruni, D., et al., "SystemC cosimulation and emulation of multiprocessor SoC designs", *IEEE Computer*, **36 (4)**, 2003.

[33] Beneš, V.E., "On rearrangeable three-stage connecting networks", *Bell Syst. Tech. J.*, **16 (5)**, 1962, pp. 1481–1491.

[34] Bermond, J.C., Comellas, F., and Hsu, D.F., "Distributed loop computer networks: a survey", *J. Parallel Distrib. Comput.*, **24 (1)**, 1995, pp. 2–10.

[35] Bertozzi, D., Benini, L., and De Micheli, G., "Low power error resilient for on-chip data buses", *in Proc. Design Automation and Test in Europe Conf.*, 2002, pp. 102–109

[36] Bertozzi, D., Benini, L., and De Micheli, G., "Error control schemes for on-chip interconnection networks: reliability versus energy efficiency", in Jantsch, A. and Tenhunen, H. (eds), *Networks on Chip*, Kluwer Academic Publisher, 2003.

[37] Bertsekas, D.P. and Gallager, R.G., *Data Networks*, Prentice Hall, 1992.

[38] Best, E., "Semantics of Sequential and Parallel Programs", Prentice Hall, 1996.

[39] Bjerregaard, T. and Sparsoe, J., "A router architecture for connection-oriented service guarantees in the Mango clockless network-on-chip", *in Proc. Design Automation and Test in Europe Conf.*, 2005.

[40] Bjerregaard, T., "The Mango clockless network-on-chip: concepts and implementation", *Ph.D. Thesis*, Computer Science, Technical University of Denmark, 2005.

[41] Bjerregaard, T. and Sparsoe, J., "A scheduling discipline for guaranteeing bandwidth and latency in asynchronous network-on-chip", *in Proc. Int. IEEE Symp. Asynch. Circ. and Syst.*, 2005.

[42] Blelloch, G.E., "NESL: a nested data parallel language", *Tech. Rep. CMU-CS-93-129*, Dept. Electr. Comp. Engin., Carnegie Mellon University, 1993.

[43] Bolotin, E., Cidon, I., Ginosar, R., and Kolodny, A., "QNoC: QoS architecture and design process for network on chip", *J. Syst. Arch.*, **50 (2-3)**, 2004, pp. 105–128.

[44] Bolsens, I., De Man H.J., Lin, B., et al., "Hardware/software co-design of digital communication systems", *in Proc. IEEE*, **85 (3)**, 1997, pp. 391–418.

[45] Borkar, S., "Intel says software developers should follow Moore's law", *DarkVision Hardware*, 11 Febr., 2008. Available from http://www.dvhardware.net/article19317.html

[46] Borkar, S., "Gigascale integration challenges and opportunities", Intel, 2007. Available from http://softwarecommunity.intel.com/articles/eng/1737.html

[47] Borkar, S., Cohn, R., Cox, G., et al., "Supporting systolic and memory communication in iWarp", *in Proc. IEEE Symp. Comp. Arch.*,1990, pp. 70–81.

[48] Boston Circuits Inc., see http://www.bostoncircuits.com

[49] Brassard, G. and Bratley, P., *Algorithmics: Theory and Practice*, Prentice-Hall, 1988.

[50] Brook, see http://graphics.stanford.edu/projects/brookgpu

[51] Brunel, J-Y., Kruijtzer, W.M., Kenter, H.J., et al. "Cosy communication IP's", *in Proc. Design Automation Conf.*, 2000, pp. 406–409.

[52] Bryant, R.E., Kwang-Ting, C., Kahng, A., et al., "Limitations and challenges of computer-aided design technology for CMOS VLSI", *in Proc. IEEE*, 89-3, 2001, pp. 341–365.

[53] Bulldast, Powerchecker: an integrated environment for RTL power estimation and optimization, version 4.0. Available from http://www.bulldast.com

[54] Bunker, A., Gopalakrishnan, G., McKee, S.A., "Formal hardware specification languages for protocol compliance verification", *ACM Trans. Design Automation of Electr. Syst.*, **9 (1)** , 2004, pp. 1–32.

[55] Burd, T., Pering, T., Stratakos, A., and Brodersen, R.W., "A dynamic voltage scaled microprocessor system", *in Proc. Int. Solid State Circ. Conf.*, 2000, pp. 294–295.

[56] Cai, L. and Gajski, D., "Transaction-level modeling in system-level design", *Tech. Rep.*, Comp. Engin. and Comp. Sci., University of California, Irvine, 2003.

[57] Caldari, M., Conti, M., Coppola, M., et al., "System-level power analysis methodology applied to the AMBA AHB bus", *in Proc. Design Automation and Test in Europe Conf.*, Vol. 20, 2003, pp. 32–39.

[58] Cantor, D.G., "On nonblocking switching networks", *Networks*, **1**, 1971, pp. 367–377.

[59] Su, R., Mittal, R., and Garg, V., "Synchronous pipelined relay stations with back-pressure tolerance", *in Proc. Int. Workshop on System-on-Chip for Real-Time Applications*, 2005, pp. 517–520.

[60] Carloni, L.P., McMillan, K.L., and Sangiovanni-Vincentelli, A.L., "Theory of latency-insensitive design", *in IEEE Trans. Comp. Aided Design of Integrated Circuits and Syst.*, **20 (9)**, 2001, pp 1059–1076.

[61] Carloni, L.P. and Sangiovanni-Vincentelli, A.L., "Coping with latency in SoC design", *IEEE Micro - Special Issue on Systems on Chip*, **22 (5)**, 2002, pp. 24–35.

[62] Chan, J. and Parameswaran, S., "NoCGEN: a template-based reuse methodology for networks on chip architecture", *in Int. Conf. VLSI Design (VLSID)*, 2004.

[63] Chandrakasan, A., Potkonjak, M., Rabaey, J., and Brodersen, R.W., "Optimizing power using transformations", *IEEE Trans. Comp. Aided Design*, **14 (1)**, 1995, pp. 12–31.

[64] Chandrakasan, A. and Brodersen, R.W., *Low Power Digital CMOS Design*, Kluwer Academic Publisher, 1995.

[65] Charest, L., Aboulhamid, E.M., and Tsikhanovich, A., "Designing with SystemC: multi-paradigm modeling and simulation performance evaluation", *in Proc. HDL Conf.*, 2001, pp. 33–45.

[66] Charest, L. Reid , M. Aboulhamid, E.M., and Bois, G., "A methodology for interfacing open source SystemC with a third party software", *in Proc. Design Automation and Test in Europe Conf.*, 2001, pp. 16–20.

[67] Chen, C. and Hwang, F.K., "The minimum distance diagram of double loop networks", *IEEE Trans. Computers*, **C-49 (9)**, 2000, pp. 977–979.

[68] Chen, T., Raghavan, R., Dale, J., and Iwata, E., "Cell broadband engine architecture and its first implementation – a performance view", *IBM J. Research and Development*, **31 (5)**, 2007. Available from http://www.research.ibm.com/journal/rd/515/chen.html

[69] Chien, A.A., Hill, M.D., and Mukherjee, S.S., "Design challenges for high-performance network interfaces: guest editor's introduction", *IEEE Computer*, **31 (11)**, 1998, pp. 42–44.

[70] Chipvision, Orinoco: a high-level power estimation and optimization tool. Available from http://www.chipvision.com

[71] Chong, F.T., Lim, B-H., Bianchini, R., et al., "Application performance on the MIT Alewife machine", *IEEE Computer*, **29 (12)**, 1996, pp. 57–64.

[72] Chung, K.M. and Wong, C.K., "Construction of a generalized connector with $5.8n \log n$ edges", *IEEE Trans. Computers*, **C-29 (11)**, 1988, pp. 1029–1032.

[73] Cierto Virtual Component Co-design (VCC), Cadence Design Systems. Available from http://www.cadence.com/technology/hwsw/ciertovcc/

[74] Cilk, see http://supertech.csail.mit.edu/cilk/

[75] Clos, C., "A study of nonblocking switching networks", *Bell Techn. J.*, 1953, pp. 407–424.

[76] Cohen, T., Sriram, N., Leland, D., et al., "Soft error considerations for deep-submicron CMOS circuit applications", *in Proc. Int. IEEE Electron Device Meeting*, 1999, pp. 315–318.

[77] Computer History Museum, see http://www.computerhistory.org

[78] Connex, see http://www.connex-electronics.com

[79] Conti, M., Pieralisi, L., Caldari, M., et al., "Power analysis methodology and library in SystemC", *in Proc. SPIE Int. Conf. VLSI Circ. Syst.*, **5837 (1)**, 2005, pp. 446–455.

[80] Coppola, M., Curaba, S., Grammatikakis, M., et al., "The OCCN user manual". Available from the ancillary book CD or http://occn.sourceforge.net

[81] Coppola, M., Curaba, S., Grammatikakis, M.D., and Maruccia, G., "IPSIM: SystemC 3.0 enhancements for refinement", *in Proc. Design Automation and Test in Europe Conf.*, 2003, pp 106–111.

[82] Coppola, M., Curaba, S., Grammatikakis, M.D., et al., "OCCN: a NoC modeling framework for design exploration", *J. Syst. Arch. - Special Issue on Network-on-Chip*, Elsevier (North Holland), **50**, 2004, pp. 129–163.

[83] Coppola, M., Pistritto, C., Locatelli, R., and Scandurra, A., "STNoC: an evolution towards MPSoC era", NoC Workshop, 2006.

[84] CoWare ConvergenSC System Designer, CoWare Inc. Available from http://www.coware.com

[85] Culler, D.E., Singh, J.P., and Gupta, A., *Parallel Computer Architecture: a Hardware/Software Approach*, Morgan-Kaufmann, 1998.

[86] Dally, W.J. and Seitz, C., "Deadlock free message routing in multiprocessor interconnection networks", *IEEE Trans. Computers*, **C-36 (5)**, 1987, pp. 547–553.

[87] Dally, W.J., "Performance analysis of k-ary n-cube interconnection networks", *IEEE Trans. Computers*, **C-39 (6)**, 1990, pp. 775–785.

[88] Dally, W.J. and Poulton, J.W., *Digital Systems Engineering*, Cambridge University Press, 1998.

[89] Dally, W.J. and Towles, B., "Route packets, not wires: on-chip interconnection networks", *in Proc. IEEE Design Automation Conf.*, 2001 pp. 684–689.

[90] Davis, A. and Nowick, S.M., "An introduction to asynchronous circuit design", *Tech. Rep. UUCS-97-013*, Dept. CS, University of Utah, 1997.

[91] Day, J. and Zimmermman, H., "The OSI reference model", *in Proc. IEEE*, 71 (12), 1983, pp. 1334–1340.

[92] de Bruijn, N.G., "A combinatorial problem", *in Proc. Abademe Van Waterschappen*, **49**, 1946, pp. 758–764.

[93] Densmore, D., "Metropolis architecture refinement styles and methodology", *Techn. Memorandum UCB/ERL M04/36*, 2004, Dept. Electr. Engin. Comp. Sci., University of California, Berkeley.

[94] Diefendorff, K. and Dubey, P., "How multimedia workloads will change processor design", *IEEE Computer*, **30 (9)**, 1997, pp. 43–45.

[95] Diefendorff, K., and Duquesne, Y., "New degress of parallelism in SoCs", *EE Times*, Sept. 13, 2002.

[96] Dielissen, J., Radulescu, A., Goossens, K., and Rijpkema, E., "Concepts and Implementation of the Philips Network-on-Chip", *in Proc. IP-Based SOC Design*, 2003.

[97] Diep, T.A. and Shen, J.R., "A visual-based microarchitecture testbench", *IEEE Computer*, **28 (12)**, 1995, pp. 57–64.

[98] Ding, J. and Bhuyan, L.N., "Performance evaluation of multistage networks with finite buffers", *in Proc. Int. IEEE Conf. Par. Proc.*, Vol. 1, 1992, pp. 592–599.

[99] Ding, J. and Bhuyan, L.N., "Finite buffer analysis of multistage interconnection networks", *IEEE Trans. Computers*, **C-43 (2)**, 1994, pp. 243–247.

[100] Ding, J. and Bhuyan, L.N., "Analysis of multi-queue buffer allocation schemes in multistage interconnection networks", *Tech. Rep. - 053/93*, Dept. CS, Texas A & M University, 1993.

[101] Dixit, A., "Networking applications for Xtensa configurable processors", *Linley Group Tech Seminars*, 2006. Available from http://www.tensilica.com/pdf/LinleyNPUSem_2007.pdf

[102] Dobravec, T., Robic, B., and Zerovnik, J., "Permutation routing in double loop networks: design and empirical evaluation", *J. Syst. Arch.*, Elsevier (North Holland), **48**, 2003, pp. 387–402.

[103] Dongarra, J., Foster, I., Fox, G., et al., *The Source Book of Parallel Computing*, Morgan-Kaufmann, 2003.

[104] Duato, J., Yalamanchili, S., and Ni, L., *Interconnection Networks, an Engineering Approach*, Morgan-Kaufmann, 2003.

[105] Duller, A., Towner, D., Panesar, G., et al., "picoArray technology: the tool's story", *in Proc. Conf. Design, Automation and Test in Europe*, Vol. 3, 2005, pp. 106–111.

[106] Dumitras, T., Kerner, S., and Marculescu, R., "Towards on-chip fault-tolerant communication", *in Proc. Asia and South Pacific Design Automation Conf.*, 2003.

[107] EDA Roadmap, www.medeaplus.org/roadmap

[108] Edman, and Svensson, C., "Timing closure through globally synchronous, timing portioned design methodology", *in Proc. Asia and South Pacific Design Automation Conf.*, 2004, pp. 71–74.

[109] Felicijan, T., Bainbridge, W.J., and Furber, S.B., "An asynchronous low latency arbiter for quality of service (QoS) applications", *in Proc. Int. Conf. Microelectronics*, 2003.

[110] Felicijan, T. and Furber, S.B., "An asynchronous on-chip network router with quality-of-service (QoS) support", *in Proc. Int. IEEE SoC Conf.*, 2004, pp. 274–277.

[111] Ferrari, A. and Sangiovanni-Vincentelli, A., "System design: traditional concepts and new paradigms", *in Proc. Conf. Computer Design*, 1999, pp. 2–13.

[112] Fidge C.J., "Partial orders for parallel debugging", *in Proc. ACM Workshop Parallel Distr. Debug.*, 1988, pp. 183–194.

[113] Forsell, M., "A scalable high-performance computing solution for networks on chips", *IEEE Micro*, **22 (5)**, 2002, pp. 46–55.

[114] 4More, http://4more.av.it.pt

[115] Furber, S.B., Efthymiou, A., and Singh, M., "A power-efficient duplex communication system", *in Proc. Int. Workshop on Asynchr. Interfaces*, 2000.

[116] Gajski, D.D., Zhu, J., Doemer, A., et al., "SpecC: specification language and methodology", Kluwer Academic Publisher, 2000. Also, see http://www.specc.org

[117] Garg, S. and Lajolo, M., "UML modeling and configuration of tile-based networks on chip", *in Proc. Int. Workshop on UML for SoC Design*, 2006, pp. 1–4.

[118] Garg, S. and Lajolo, M., "C-based design of a flexible wrapper for tiled networks on chip", *in Proc. Forum on Specification and Design Languages (FDL)*, 2006, pp. 185–188.

[119] Gebremichael, B., Vaandrager, F.W., Zhang, M., et al., "Deadlock prevention in the Æthereal protocol", *in Proc. Int. Conf. Chips for Application-Specific Systems on Chips (CHARME)*, 2005, pp. 345–348.

[120] Genko, N., Atienza, D., De Micheli, G., et al., "A novel approach to NoC Emulation", *in Proc. Int. Symp. Circ. Syst.*, 2005, pp. 2365–2368.

[121] Gerstlauer, A., Yu, H., and Gajski, D., "RTOS modeling for system level design", *in Proc. Design Automation and Test in Europe Conf.*, 2003, pp. 130–135.

[122] Givargis, T., Vahid, F., and Henkel, J., "Instruction based system level power evaluation of system-on-a-chip peripheral cores", *IEEE Trans. VLSI Syst.*, **10 (6)**, 2002, pp. 856–863.

[123] Goke, L.R. and Lipovski, G.J., "Banyan networks for partitioning multiprocessor systems", *in Proc. 1st Annual Symp. on Comp. Architecture*, 1973, pp. 21–28.

[124] Goossens, K., Radulescu, A., Wielage, P., et al., "Networks on silicon: combining best-effort and guaranteed services", *in Proc. Design Automation and Test in Europe Conf.*, 2002, pp. 423–427.

[125] Goossens, K., Dielissen, J., van Meerbergen, J., et al., "Guarranteeing the quality of services in networks on chip", in Jantsch, A. and Tenhunen, H. (eds), *Networks on Chip*, Kluwer Academic Publisher, Chapter 4, 2003, pp. 61–82.

[126] Goossens, K., Radulescu, A., van Meerbergen, J., et al., "Tradeoffs in the design of a router with both guaranteed and best-effort services for networks on chip", *in Proc. Design Automation and Test in Europe Conf.*, 2003, pp. 350–355.

[127] Gordon Moore's law, see ftp://download.intel.com/museum/Moores_Law

[128] Le Gourrierec, M., "System and method for providing multiple quality of service levels over a single asynchronous transfer mode virtual communications channel", *Int. Patent EP1209864* , Sagem Tech., 2002. Available from http://www.freepatentsonline.com/H002051.html

[129] Grammatikakis, M.D., Hsu, D.F., and Kraetzl, M., *Parallel System Interconnections and Communications*, CRC Press, 2000.

[130] Grammatikakis, M.D., and Liesche, S. "Priority queues and sorting for parallel simulation", *IEEE Trans. Soft. Engin.*, **SE-26 (5)**, 2000, pp. 401–422.

[131] Greenstreet, M.R., "Implementing a STARI chip", *Int. Conf. Comp. Design*, 1995.

[132] Guerrier, P., and Greiner, A., "A generic architecture for on-chip packet-switched interconnections", *in Proc. Design Automation and Test in Europe Conf.*, 2000, pp. 250–256.

[133] Guz, Z., Walter, I., Bolotin, E., Cidon, I., et al., "Efficient link capacity and QoS design for network-on-chip", *in Proc. Design Automation and Test in Europe Conf.*, 2006, pp. 9–14.

[134] Hammond, L., Carlstrom, B.D., Wong, V., et al., "Transactional coherence and consistency: simplifying parallel hardware and software", *IEEE Micro*, **24 (6)**, pp. 92–103.

[135] Harel, D., "Statecharts: a visual formalism for complex systems", *Sci. Comp. Programming*, **8**, 1987, pp. 231–274.

[136] Harper, D.T. and Jump, J.R., "Evaluation of reduced bandwidth multistage networks", *J. Parallel Distr. Comput.*, **C-9**, 1990, pp. 304–311.

[137] Haverinen, A., Leclercq, M., Weyrich, N., et al., "SystemC-based SoC communication modeling for the OCP protocol", *White Paper* , OCP-IP, 2002. Available from http://www.ocpip.org

[138] Hedge, R. and Shanbhag, N., "Towards achieving energy efficiency in presence of deep submicron noise", *IEEE Trans. VLSI Syst.*, **8 (4)**, 2000, pp. 379–391.

[139] Hemani, A., Jantsch, A., Kumar, S., et al., "Network on a chip: an architecture for billion transistor era", *in Proc. Int. IEEE Norchip Conf.*, 2002.

[140] Henkel, J. and Li, Y., "Avalanche: an environment for design space exploration and optimization of low-power embedded systems", *IEEE Trans. VLSI Integr. Syst.*, **10 (4)**, 2002, pp. 454–468.

[141] Henry, D.S. and Joerg, C.F., "A tightly-coupled processor-network interface", *in Proc. Int. Conf. Arch. Support Progr. Lang. and Oper. Syst.*, 1992.

[142] Herlihy, M.P., "Impossibility and universality results for wait-free synchronization", *in Proc. Int. Conf. Princ. Distr. Comp. (PODC)*, 1988, pp. 276–290.

[143] Herlihy, M., and Moss, J.E.B., "Transactional memory: architectural support for lock-free data structures", *in Proc. Int. IEEE Symp. Comp. Arch.*, 1993, pp. 289–300.

[144] Herlihy, M., "A methodology for implementing highly concurrent data objects", *IEEE Trans. Progr. Lang. and Syst.*, **15 (5)**, 1993, pp. 745–770.

[145] Herlihy, M.P., Luchangco, V., and Moir, M., "Obstruction-free synchronization: double-ended queues as an example", *in Proc. Int. Conf. Distr. Comp. Syst. (ICDCS)*, 2003, pp. 522–529.

[146] Herlihy, M., Luchangco, V., Moir, M., Scherer, W.N., et al., "Software transactional memory for dynamic-sized data structures", *in Proc. Int. Conf. Distr. Comp. Syst. (ICDCS)*, 2003, pp. 92–101.

[147] Heydemann, M.C., Opatrny, J., and Sotteau, D., "Broadcasting and spanning trees in de Bruijn and Kautz networks", *Discr. Appl. Math.*, **37 (38)**, 1992, pp. 297–317.

[148] Hill, M.D., "Multiprocessors should support simple memory consistency models", *IEEE Computer*, **C-31 (8)**, 1998, pp. 28–34.

[149] Hluchyj, M.G. and Karol, M.J., "Queuing in high-performance packet switching", *IEEE Trans. Sel. Areas Comm.*, **C-6 (9)**, 1988, pp. 1587–1597.

[150] Ho, W.H. and Pinkston, T.M., "A methodology for designing efficient on-chip interconnects on well-behaved communication patterns", *in Proc. Int. Symp. High-Performance Computer Architecture*, 2003, pp. 377–388.

[151] Hoare, R.R., Ding, Z., Tung, S., et al., "A framework for the design, synthesis and cycle-accurate simulation of multiprocessor networks", *J. Parallel Distrib. Comput.*, **65 (10)** , 2005, pp. 1237–1252.

[152] Holzmann, G.J., *Design and Validation of Computer Protocols*, Prentice Hall, 1991.

[153] Hsu, D.F. and Wei, D.S., "Efficient routing and sorting schemes for de Bruijn networks", *IEEE Trans. Parallel Distrib. Syst.*, **8**, 1997, pp. 1–14.

[154] Hu, J. and Marculescu, R., "Energy-performance aware mapping for regular NoC architectures", *IEEE Trans. Comp. Aided Design Integr. Circ. Syst.*, **24 (4)**, 2005, pp. 551–562.

[155] Hwang, F.K., "A permutation routing algorithm for double loop networks", *Parallel Proc. Letters*, **7 (3)**, 1997, pp. 259–265.

[156] Hwang, F.K., "A complementary survey on double-loop networks", *Theor. Comput. Sci.*, **263 (1-2)**, 2001, pp. 211–229.

[157] Hwang, F.K., "A survey on multi-loop networks", *Theor. Comput. Sci.*, Elsevier (North Holland), **299 (1-3)**, 2003, pp. 107–121.

[158] IBM, http://www.ibm.com

[159] IBM Blue Logic Technology, see http://www-3.ibm.com/chips/bluelogic/

[160] IBM CoreConnect, see http://www.chips.ibm.com/products/coreconnect

[161] Iliadis, I. and Denzel, W.E., "Analysis of packet switches with input and output queuing", *IEEE Trans. Comm.*, **C-41 (5)**, 1993, pp. 731–740.

[162] Imagine, see http://cva.stanford.edu/projects/imagine/

[163] Intel, see http://www.intel.com

[164] Intellasys, see http://www.intellasys.net/products/seaforth

[165] International Business Solutions, "Global system IC ASSP/ASIC service management report", 2007.

[166] IRAM, see http://iram.cs.berkeley.edu

[167] ITRS, see http://public.itrs.net

[168] Jalabert, A., Murali, S., Benini, L., and De Micheli, G., "xpipesCompiler: a tool for instantiating application specific networks on chip", *in Proc. Design, Automation and Test in Europe Conf.*, 2004, pp. 884–889.

[169] Jerraya, A. and Wolf, W., *Multiprocessor Systems-on-Chip*, Morgan-Kauffman, 2004.

[170] Jesshope, C.R., Miller, P.R., and Yantchev, Y.S., "High performance communications in processor networks", *in Proc. Int. IEEE Symp. Comp. Arch.*, 1989, pp. 150–157.

[171] Jesshope, C.R., Miller, P.R., and Yantchev, Y.S., "The mad-postman network chip", *in Proc. Transp. Conf.*, Vol. II, 1991, pp. 517–536.

[172] Jhaveri, T., Pillegi, L., Rovner, V., et al., "Maximization of layout printability/manufacturability by extreme layout regularity", *in Proc. SPIE Microlithography*, Vol. 6156, 2006.

[173] Jhaveri, T., Rovner, V., Pillegi, L., et al., "Maximization of layout printability/ manufacturability by extreme layout regularity", *J. Micro/Nanolithography*, **6 (3)**, 2007, pp. 1–15.

[174] Kahng, A.B., "Design challenges at 65nm and beyond", *in Proc. Design Automation and Test in Europe Conf.*, 2007, 1466–1467.

[175] Kapasi, U.J., Rixner, S., Dally, W.J., et al., "Programmable stream processors", *IEEE Computer*, **36 (8)**, 2003, pp. 54–62.

[176] Kaplan, A., Sarrafzadeh, M., and Kastner, R., "A survey of hardware/software system partitioning", *Tech. Rep.*, Comp. Sci., University of California, Santa Barbara, 2003.

[177] Karim, F., Nguyen, A., and Dey, S., "An interconnect architecture for network systems on chips", *IEEE Micro*, **22 (5)**, 2002, pp. 36–45.

[178] Karim, F., Nguyen, A., Dey, S., and Rao, R., "On-chip communication architecture for OC-768 network processors", *in Proc. Asia and South Pacific Design Automation Conf.*, 2001, pp. 678–683.

[179] Katevenis, M., Vatsolaki, P., and Efthymiou, A., "A pipelined memory shared buffer for VLSI switches", *in Proc. ACM SIG Data Communication*, 1995, pp. 39–48.

[180] Kessler, R.E., Thorson, G., Birritella, M., Oberlin, S., and Passint, R., "Multiprocessor computer system with interleaved processing element nodes", *Int. Patent US5737628*, Cray Research Inc., 1998.

[181] Keutzer, K., Malik, S., Newton, A.R., et al., "System-level design: orthogonalization of concerns and platform-based design", *in IEEE Trans. Comp. Aided Design of Integr. Circ. and Systems*, **19 (12)**, 2000, pp. 1523–1543.

[182] Kheterpal, V., Rovner, V., Hersan, T.G., et al., "Design methodology for IC manufacturability based on regular logic bricks", in *Proc. Design Automation Conf.*, 2005, pp. 353–358.

[183] Kim, J., Balfour, J., and Dally, W.J., "Flattened butterfly topology for on-chip networks", *Int. Symp. Microarchitecture (MICRO)*, 2007.

[184] Kim, J., Dally, W.J., and Abts, D., "Flattened butterfly: a cost-efficient topology for high-radix networks, *in Proc. Int. IEEE Symp. Comp. Arch.*, 2007, pp. 126–137.

[185] Kim, S. and Sridhar, R., "Self-timed mesochronous interconnections for high-speed VLSI systems", *in Proc. Great Lake Symp. on VLSI (GLSVLSI)*, 1996, pp. 122–128.

[186] Kim, N.S., Austin, T., Baauw, D., et al., "Leakage current: Moore's law meets static power", *IEEE Computer*, **36 (12)**, 2003, pp. 68–75.

[187] Klasing, R., "The relationship between the gossip complexity in vertex-disjoint paths mode and the vertex bisection width", *Discr. Appl. Math.*, **83 (1)**, 1998, pp. 229–246.

[188] Kogel, T., Doerper, M., Wieferink, A., et al., "A modular simulation framework for architectural exploration of on-chip interconnection networks", *in Int. Conf. Hardware/Software Codesign and System Synthesis (CODES+ISSS)*, 2003, pp. 7–12.

[189] Koutsofios, E. and North, S.C., "Dot User Manual", *Techn. Rep.* , AT&T Bell Labs, Murray Hill, NJ, 1993. See http://www.cs.brown.edu/cgc/papers/dglpttvv-ddges-97.ps.gz

[190] Krolak, D., "Unleashing the Cell broadband engine processor", *MPR Fall Processor Forum*, 2005.

[191] Krolikoski, S., Schirrmeister, F., Salefski, B., et al., "Methodology and technology for virtual component driven hardware/software co-design on the system level", *in Proc. Int. IEEE Symp. Circ. and Syst.*, 1999, pp. 456–459.

[192] Kubiatowicz, J.D., "Integrated message-passing and shared-memory communication in the Alewife multiprocessor", *Ph.D. Thesis*, MIT, Dept. Electr. Engin. Comp. Sci., 1998.

[193] Kumar, V., Grama, A., Gupta, A., and Karypis, G., *Introduction to Parallel Computing*, Benjamin Cummings, 1994.

[194] Kumar, N., Katkoori, S., Rader, L., and Vemuri, R., "Profile-driven behavioral synthesis for low power VLSI systems", *IEEE Design & Test of Computers*, **12 (3)**, 1995, pp. 70–84.

[195] Kumar, S., Jantsch, A., Soininen, J., et al., "A network on chip architecture and design methodology", *in Proc. Int. Symp. VLSI*, 2002, pp. 117–124.

[196] Labonte, F., Mattson, P.R., Thies, W., et al., "The Stream virtual machine", *in Proc. Int. Conf. Parallel Arch. and Compiler Tech. (PACT)*, 2004, pp. 267–277.

[197] Lahiri, K., Raghunathan, A., and Dey, S., "Design space exploration for optimizing on-chip communication networks", *in IEEE Trans. Comp. Aided Design of Integr. Circ. and Syst.*, **23 (6)**, 2004, pp. 952–961.

[198] Lahiri, K., Raghunathan, A., and Dey, S., "System level performance analysis for designing on-chip communication architectures", *in IEEE Trans. Comp. Aided Design of Integr. Circ. and Syst.*, **20 (6)**, 2001, pp. 768–783.

[199] Lahiri, K., Raghunathan, A., and Dey, S., "Evaluation of the traffic performance characteristics of SoC communication architectures", *in Proc. Conf. VLSI Design*, 2001, pp. 29–34.

[200] Lajolo, T.M., Raghunathan, A., Dey, S., and Lavagno, L., "Efficient power co-estimation techniques for system-on- chip design", *in Proc. Design Automation and Test in Europe Conf.*, 2000.

[201] Lamport, L., "Concurrent reading and writing", *Comm. ACM*, **20 (11)**, 1977, pp. 806–811.

[202] Lamport, L., "How to make a multiprocessor computer that correctly executes multiprocess programs", *IEEE Trans. Computers*, **C-28 (9)**, 1979, pp. 690–691.

[203] Laudon, J. and Lenoski, D., "The SGI Origin: a ccNUMA highly scalable server", *in Proc. Int. Symp. on Comp. Arch.*, 1997.

[204] Lawrie, D.H., "Access and alignment of data in an array processor", *IEEE Trans. Computers*, **C-24 (12)**, 1975, pp. 1145–1155.

[205] Lea, C.T. and Shyy, D.J., "Tradeoff of horizontal decomposition versus vertical stacking in rearrangeable nonblocking networks", *IEEE Trans. Comm.*, **39 (6)**, 1991, pp. 899–904.

[206] Lee, K., Lee, S.-J., and Yoo, H.-J., "Low-power network-on-chip for high-performance SoC design", *IEEE Trans. VLSI Syst.*, **14 (2)**, 2006, pp. 148–160.

[207] Lee, E.A., "Overview of the Ptolemy Project", *Techn. Memo UCB/ERL M01/11*, 2001, Dept. Electr. Engin. Comp. Sci., University of California, Berkeley. Available from http:/ptolemy.eecs.berkeley.edu/index

[208] Leighton, F.T., *Introduction to Parallel Algorithms and Architectures*, Academic Press, 1992.

[209] Leiserson, C.E., Abuhamdeh, Z.S., Douglas, D.C., et al., "The network architecture of the Connection Machine CM-5", *J. Parallel Distrib. Comput.*, **33 (2)**, 1996, pp. 145–158.

[210] Leiserson, C.L., "Fat trees: universal networks for hardware-efficient supercomputing", *IEEE Trans. Computers*, **C-34 (10)**, 1985, pp. 892–901.

[211] Liang, J., Swaminathan, S., and Tessier, R., "aSOC: a scalable, single-chip communication architecture", *in Int. IEEE Conf. Parallel Arch. and Compil. Techn.*, 2000, pp. 37–46.

[212] LibSh, see http://www.libsh.org

[213] LibSimd, see http://libsimd.sourceforge.net

[214] Liebmann, L., Barish, A., Baum, Z., et al., "High-performance circuit design for the RET-enabled 65nm technology node", in Liebmann, L.W. (ed), *Proc. SPIE, Design and Process Integration for Microelectronic Manufacturing II*, **5379**, 2004, pp. 20–29.

[215] Liestman, A.L., Opartny, J., and Zaragoza, M., "Network properties of double and triple fixed step graphs", *Int. J. Found. Comp. Sci.*, **9**, 1998, pp. 57–76.

[216] Limo Foundation Platform, see http://www.limofoundation.org/images/stories/pdf/limo_platform_arch.pdf

[217] Lin, A. and Pippenger, N., "Parallel algorithms for routing in nonblocking networks", *Math. Syst. Theory*, **27 (1)**, 1994, pp. 29–40.

[218] Lines, A., "Asynchronous interconnect for synchronous SoC design", *IEEE Micro*, **24**, 2004, pp. 32–41.

[219] Lisinski, D., Leighton, F.T., and Maggs, B.M., "Empirical evaluation of randomly-wired multistage networks", *in Proc. Int. IEEE Conf. Comput. Design.*, 1998, pp. 380–385.

[220] Liu, X. and Papaefthymiou, M.C., "HyPE: hybrid power estimation for ip-based programmable systems", *in Proc. Asia and South Pacific Design Automation Conf.*, 2003, pp. 606–609.

[221] Mackenzie, K.M., "An efficient virtual network interface in the FUGU scalable workstation", *Ph.D. Thesis*, MIT, Dept. Electr. Engin. Comp. Sci., 1998.

[222] Mackenzie, K., Kubiatowicz, J., Agarwal, A., and Kaashoek, M.F., "FUGU: implementing translation and protection in a multiuser, multimodel multiprocessor", *Tech. Memo TM-503*, MIT, LCS, 1994.

[223] Maggs, B.M., "Randomly wired multistage networks", *Statistical Sci.*, **8 (1)**, 1993, pp. 70–75.

[224] Market research, "Consumer electronic processors: strategic assessment of key trends and market drivers through 2025", see http://www.marketresearch.com/map/prod/1305700.html

[225] Massalin, H. and Pu, C., "A lock-free multiprocessor OS kernel", *Tech. Rep. CUCS-005-91*, Columbia Univ., 1991.

[226] Medvidovic, N. and Taylor, R.N., "A framework for classifying and comparing architecture description languages", *in Proc. Softw. Engin. Conf.*, LNCS 1301, 1997, pp. 60–76.

[227] Mehta, H., Owens, R., Irwin, M., et al., "Techniques for low energy software", *Int. Symp. Low Power Electr. and Design*, 1997, pp. 72–75.

[228] Melkonian, M., "Get by without an RTOS", *Embedded Syst.*, **13 (10)**, Sept. 2000. Available from http://www.embedded.com/2000/0009/0009feat4.htm

[229] Mello, A., Palma, J., Moraes, F., and Calazans, N., "MAIA: a framework for networks on chip generation and verification", *in Proc. Asia and South Pacific Design Automation Conf.*, 2005, pp. 49–52.

[230] Mellor-Crummey, J.M. and Scott, M.L., "Algorithms for scalable synchronization on shared-memory multiprocessors", *ACM Trans. Comp. Syst.*, **C-9 (1)**, 1991, pp. 21–65.

[231] Merritt, R., "Intel developing languages, eyes multimedia applications", *EE Times*, May 4, 2007.

[232] Mesgarzadeh, B., Svensson, C., and Alvandpour, A., "A new mesochronous clocking scheme for synchronization in SoC", *in Proc Symp. Circuits and Syst*, 2002, pp. 605–609.

[233] Metropolis project, see http://www.gigascale.org/metropolis

[234] Millberg, M., Nilsson, E., Thid, R., and Jantsch, A., "The Nostrum backbone - a communication protocol stack for networks on chip", *in Proc. Int. Conf. VLSI Design*, 2004.

[235] Millberg, M., Nilsson, E., Thid, R., and Jantsch, A., "Guaranteed bandwidth using looped containers in temporally disjoint networks within the

Nostrum network on chip", *in Proc. Design, Automation and Test in Europe Conf.*, 2004.

[236] Monchiero, M., Palermo, G., Silvano, C., and Villa, O., "Exploration of distributed shared memory architectures for NoC-based multiprocessors", *J. Syst. Arch.*, **53 (10)**, 2007, pp. 719–732.

[237] Moraes, F.G., Calazans, N., Mello, A., et al., "HERMES: an infrastructure for low area overhead packet-switching networks on chip ", *J. VLSI Integration*, **38 (1)**, 2004, pp. 69–93.

[238] Moraes, F.G., Mello, A., Moeller, L., et al., "A low area overhead packet-switched network on chip: architecture and prototyping", *in Proc. Int. Conf. VLSI*, 2003, pp. 318–323.

[239] Moriconi, M., Qian, X., and Riemenschneider, R.A., "Correct architecture refinement", *IEEE Trans. Softw. Engin.*, **SE-21 (4)**, 1995, pp. 356–372.

[240] Mu, F. and Svensson, C., "Self-tested self-synchronization circuit for mesochronous clocking", *IEEE Trans. Circ. and Syst. II: Analog and Digital Signal Proc.*, **48 (2)**, 2001, pp. 129–141.

[241] Mueller, W., Ruf, J., Hoffmann, D., et al., "The simulation semantics of SystemC", *in Proc. Design Automation and Test in Europe Conf.*, 2001, pp. 64–71.

[242] Mukherjee, B., *Optical Communication Networks*, McGraw-Hill, 1997.

[243] Mukherjee, S.S., Falsafi, B., Hill, M.D., and Wood, D.A., "Coherent network interfaces for fine-grain communication", *in Proc. IEEE Symp. Comp. Arch.*, 1996, pp. 247–258.

[244] Mukherjee, S. and Hill, M.D., "Making network interfaces less peripheral", *IEEE Computer*, **31 (10)**, 1998, pp. 70–76.

[245] Mukherjee, S.S., Silla, F., Bannon, P., Emer, J., et al., "A comparative study of arbitration algorithms for the Alpha 21364 pipelined router", *in Proc. Int. Conf. Arch. Support for Progr. Lang. Oper. Syst.*, 2002, pp. 223–234.

[246] Murali, S. and De Micheli, G., "Bandwidth-constrained mapping of cores onto NoC architectures", *in Proc. Design, Automation and Test in Europe Conf.*, 2004.

[247] Murali, S. and De Micheli, G., "SUNMAP: a tool for automatic topology selection and generation for NoC", *in Proc. Design Automation Conf.*, 2004.

[248] Murali, S., Meloni, P., Angiolini, F., et al., "Designing message-dependent deadlock free networks on chips for application-specific systems on chips", *in Proc. Int. Conf. VLSI*, 2006, pp. 158–163.

[249] Mutz, S. and Durieux, P., "Heterogeneous multiprocessing for efficient multi-standard high definition video decoding", *in Proc. Symp. Hot Chips*, 2006.

[250] Multicore Organization, see http://www.multicore.org

[251] Jantsch, A. and Tenhunen, H. (eds), *Networks on Chip*, Kluwer Academic Publisher, 2003.

[252] Nardi, A. and Sangiovanni-Vincentelli, A.L., "Logic synthesis for manufacturability", *IEEE Design and Test of Computers*, **21**, 2004, pp. 192–199.

[253] Nilsson, E. and Oeberg, J., "Reducing power and latency in 2D mesh NoCs using globally pseudochronous locally synchronous clocking", *in Proc. ISSS'04*, 2004, pp. 176–182.

[254] Nomadik, internal report, ST Microelectronics, 2007. Available from http://www.st.com/stonline/books/pdf/docs/9306.pdf

[255] OCP-IP. Open Core Protocol Specification. Available from http://www.ocpip.org

[256] Ofman, J.P., "A universal automaton", *Trans. Moscow Math Soc.*, 1965, pp. 200–215.

[257] Oliker, L. and Biswas, R., "Parallelization of a dynamic unstructured application using three leading paradigms", *in Proc. Int. Conf. Supercomput.*, 1999, pp. 11–18.

[258] Olukotun, K., Nayfeh, B.A., Hammond, L., et al., "The case for a single-chip multiprocessor", *in Proc. Int. Symp. Arch. Support for Parallel Lang. and Oper. Syst. (ASPLOS)*, 1996.

[259] Open Microprocessor Systems Initiative, "PI Bus draft standard specification", *Techn. Rep.*, 1994.

[260] OPNET, see http://www.opnet.com

[261] Osterling, A., Benner, T., Ernst, R., et al., "The COSYMA system", in Balarin, F., Chiodo, N., Giusto, P., et al. (eds), *Hardware/Software Co-Design: Principles and Practice*, Kluwer Academic Publisher, Chapter 8, 1997.

[262] Palmchip, "Overview of CoreFrame architecture", *White Paper*, 2002. Available from http://www.palmchip.com

[263] Pande, R.P., Grecu, C., Ivanov, A., and Saleh, R., "Design of a switch for network on chip applications", *in Proc. Int. Symp. Circ. Syst.*, 2003, pp. 217–220.

[264] Parhami, B. and Kwai, D.-M., "Periodically regular chordal rings", *IEEE Trans. Parallel Distr. Syst.*, **10 (6)**, 1999, pp. 658–672.

[265] Parker, D.S., "Notes on shuffle/exchange-type switching networks", *IEEE Trans. Computers*, **29 (4)**, 1980, pp. 213–222.

[266] Pasricha, S., "Transaction level modeling of SoC with SystemC 2.0", *in Synopsys Users Group Conf. (SNUG)*, 2002.

[267] Patel, J.H., "Processor-memory interconnections for multiprocessors", *in Proc. 6th Annual Symp. on Computer Architecture*, 1979, pp. 168–177.

[268] Paulin, P., Pilkington, C., and Bensoudane, E., "StepNP: a system-level exploration platform for network processors", *in IEEE Design and Test*, **19 (6)**, 2002, pp. 17–26.

[269] Pavia, J.M., "Design and implementation of a fat tree network on chip", *M.S. Thesis*, Inst. Microelectronics and Info. Tech., Royal Inst. Techn. (KTH), Stockholm, Sweden, 2004.

[270] Pestana, S.G., Rijpkema, E., Radulescu, A., et al., "Cost-performance trade-offs in networks on chip: a simulation-based approach", *in Proc. Design Automation and Test in Europe Conf.*, 2004, pp. 764–769.

[271] Pieralisi, L., *Ph.D. Thesis*, TIMA, Grenoble, France. Available from http://tima.imag.fr/publications/files/th/2006/mrc_241.pdf

[272] PIM, see http://www.nd.edu/~pim

[273] Pinto, A., Bonivento, A., Sangiovanni-Vincentelli, A., et al., "System level design paradigms: platform-based design and communication synthesis", *ACM Trans. Design Autom. Electr. Syst.*, **11 (3)**, 2006, pp. 537–563.

[274] Pippenger, N., "Communication networks", in van Leeuwen, J. (ed), *Handbook of Theoret. Comput. Sci.*, North-Holland, Amsterdam, 1990, pp. 807–833.

[275] Poursepanj, A., "The PowerPC performance modeling methodology", *Comm. ACM*, **37 (6)**, 1994, pp. 47–55.

[276] Powell, S. and Chau, P., "Estimating power dissipation of VLSI signal processing chips: the PFA technique", *in Proc. IEEE Workshop VLSI Sign. Proc.*, IV, 1990, pp. 250–259.

[277] PowerChecker, BullDast, http://www.bulldast.com/powerchecker.html

[278] Qin, W., Rajagopalan, S., Vaccharajani, M., et al., "Design tools for application specific embedded processors", *in Proc. Int. Workshop on Embedded Software (EMSOFT)*, 2002.

[279] Radulescu, A., Dielissen, J., Goossens, K., et al., "An efficient on-chip network interface offering guaranteed services, shared-memory abstraction, and flexible network programming", *in Proc. Design Automation and Test in Europe Conf.*, 2004.

[280] Radulescu, A., Dielissen, J., Pestana, S.G., et al., "An efficient on-chip network interface offering guaranteed services, shared-memory abstraction, and flexible network programming", *IEEE Trans. CAD Integr. Circ. Syst.*, **24 (1)**, 2005.

[281] Raghunathan, A. and Jha, N.K., "SCALP: an iterative-improvement-based low-power data path synthesis system", *IEEE Trans. CAD Integr. Circ. and Syst.*, **16 (11)**, 1997, pp. 1260–1277.

[282] Raghunathan, V., Srivastava, M.B., and Gupta, R.K., "A survey of techniques for energy efficient on-chip communication", *in Proc. Design Automation Conf.*, 2003.

[283] Rapidmind, see http://www.rapidmind.com

[284] Raw Architecure Workstation, see http://www.cag.lcs.mit.edu/raw

[285] Ready, J.F., "VRTX: a real-time operating system for embedded microprocessor applications", *IEEE Micro*, **6 (4)**, 1986, pp. 8–17.

[286] Realworldtech, see http://www.realworldtech.com/

[287] Reid, M., Charest, L., Tsikhanovich, A., et al., "Implementing a graphical user interface for SystemC", *in Proc. HDL Conf.*, 2001, pp. 224–231.

[288] Rijpkema, E., et al., "A router architecture for networks on silicon", *in Proc. Conf. Progress - 2nd Workshop on Embedded Systems*, 2001.

[289] Rosiello, A.P.E., Ferrandi, F., Pandini, D., and Sciuto, D., "A hash-based approach for functional regularity extraction during logic synthesis", *Int. Conf. on VLSI (ISVLSI)*, 2007, pp. 92–97.

[290] Rossi, D., Angelini, P., and Metra, C., "Error control scheme for NoC signal integrity", *in Proc. On-Line Testing Symp.*, 2007, pp. 43–48.

[291] Rowson, J.A. and Sangiovanni-Vincentelli, A.L., "Interface-based design", *in Proc. Asia and South Pacific Design Automation Conf.*, 1997, pp. 178–183.

[292] R-stream, see http://www.reservoir.com/r-stream.php

[293] Runtests Lite: User Guide, internal document, ST Microelectronics, 2000.

[294] Saad, V. and Schultz, M.H., "Topological properties of hypercubes", *IEEE Trans. Computers*, **37**, 1980, pp. 867–872.

[295] Salminen, E., Kangas, T., Hamalainen, T.D., et al., "HIBI communication network for system-on-chip", *J. VLSI Signal Proc. Syst.*, **43 (2-3)**, 2006, pp. 105–205.

[296] Salminen, T. and Soininen, J.-P., "Evaluating application mapping using network simulation", *in Proc. Symp. on System-on-Chip*, 2003.

[297] San Martin, R. and Knight, J., "Optimizing power in ASIC behavioral synthesis", *IEEE Design & Test of Computers*, **13** (**2**), 1996, pp. 58–70.

[298] Sangiovanni-Vincentelli, A. and Martin, G., "Platform-based design and software design methodology for embedded systems", *IEEE Design and Test of Comp.*, **18** (**6**), 2001, pp. 23–33.

[299] Sangiovanni-Vincentelli, A., Carloni, L.P., De Bernardinis, F., and Sgroi, M., "Benefits and challenges for platform-based design", *in Proc. Design Autom. Conf.*, 2004, pp. 409–414.

[300] Savage, S., Burrows, M., Nelson, G., et al., "Eraser: a dynamic data race detector for multi-threaded programs", *in Proc. ACM Symp. OS Princ.*, 1997, pp. 26–37.

[301] Scandurra, A., Falconeri, G., and Jego, B., "STBus communication system: concepts and definitions", internal document, ST Microelectronics, 2002.

[302] Scandurra, A., "STBus communication system: architecture specification", internal document, ST Microelectronics, 2002.

[303] Schwartz, J.T., Dewar, R.B.K., Dubinsky, E., and Schonberg, E., *Programming with Sets: an Introduction to SETL*, Springer-Verlag, New York, 1986.

[304] Scott, S., Abts, D., Kim, J., and Dally, W.J., "The BlackWidow high-radix Clos network", *in Proc. Int. IEEE Symp. Comp. Arch.*, 2006, pp. 16–28.

[305] Seamless HW/SW co-verification, platform-based SoC design and verification datasheet, Mentor Graphics. Available from http://www.mentor.com/platform_ex/platform_ex_ds.html.

[306] Seepold, R. and Madrid, N.M. (eds), *Virtual Component Design and Reuse*, Kluwer Academic Publisher, 2001.

[307] Selic, B., Gullekson, G., and Ward, P.T., *Real-Time Object-Oriented Modeling*, J. Wiley & Sons, New York, 1994.

[308] Seo, S.W. and Feng, T.Y., "A new routing scheme for concatenating two omega networks", *in Proc. Parallel Lang. Arch. Eur.*, Lect. Notes Comp. Sci., Vol. 817, 1994, pp. 831–834.

[309] Sequence Design, "Sequence unveils clock power, voltage island analysis", SoCcentral, July 21, 2006. Available from http://www.soccentral.com

[310] Shandle, J. and Martin, G., "Making embedded software reusable for SoCs", *EE Design*, March 1, 2002.

[311] Shannon, C.E., "A mathematical theory of communication", *Bell Syst. Techn. J.*, **27**, 1948, pp. 379–423 and 623–656.

[312] Shin, D., Abdi, S., and Gajski, D., "Automatic generation of bus functional models from transaction level models", *in Proc. Asia and South Pacific Design Automation Conf.*, 2004, pp. 756–759.

[313] Shin, D., Gerstlauer, A., Peng, J., et al., "Automatic generation of transaction level models for rapid design space exploration", *in Int. Conf. Hardware/Software Codesign and System Synthesis (CODES+ISSS)*, 2006, pp. 64–69.

[314] Shin, D., and Kim, J., "Intra-task voltage scheduling on DVS-enabled hard real-time systems", *IEEE Trans. Comp. Aided Design Integr. Circ. Syst.*, **24 (10)**, 2005, pp. 1530–1549.

[315] Shyy, D.J. and Lea, C.T., "$\log_2^d(n, m, p)$ strictly nonblocking networks", *IEEE Trans. Comm.*, **39 (10)**, 1991, pp. 1502–1510.

[316] Shyy, D.J. and Lea, C.T., "Rearrangeable nonblocking $\log_2^d(n, m, p)$ networks", *IEEE Trans. Comm.*, **42 (5)**, 1994, pp. 2084–2086.

[317] Silistix, see http://www.silistix.com

[318] SIMDx86, see http://simdx86.sourceforge.net

[319] Simunic, T., Benini, L., and De Micheli, G., "Cycle-accurate simulation of energy consumption in embedded systems", *in Proc. Int. IEEE Design Automation Conf.*, 1999, pp. 867–872.

[320] Sinha, V., Doucet, F., Siska, C., et al., "YAML: a tool for hardware design, visualization, and capture", *in Proc. Symp. Syst. Synthesis*, 2000, pp. 9–17.

[321] Sinha, A. and Chandrakasan, A., "JouleTrack: a web-based tool for software energy profiling", *in Proc. Design Automation Conf.*, 2001.

[322] "Skew Insensitive Link (SKIL): a mesochronous link for System-on-Chip communication", European Patent, ST Microelectronics, 2007.

[323] Smart Memories, see http://www-vlsi.stanford.edu/smart_memories/

[324] Snyder, L., "Introduction to the configurable, highly-parallel computer", *IEEE Trans. Computers*, **15 (1)**, 1982, pp. 47–58.

[325] Song, Y.H. and Pinkston, T.M., "Efficient handling of message-dependent deadlock", *in Proc. Int. Parallel and Distr. Proc. Symp. (IPDPS)* , 2001.

[326] Song, Y.H. and Pinkston, T.M., "A progressive approach to handling message-dependent deadlock in parallel computer systems", *IEEE Trans. Parallel Distr. Syst.*, **14 (3)**, 2003.

[327] Sonics Backplane Micro-Network, see http://www.sonicsinc.com

[328] Soederquist, I., "Globally updated mesochronous design style", *in IEEE J. Solid-State Circ.*, **38 (7)**, 2003, pp. 1242–1249.

[329] Split-C, see http://www.eecs.berkeley.edu/Research/Projects/CS/parallel/castle/split-c/

[330] Spirit Consortium, see http://www.spiritconsortium.org

[331] Sridhar, M.A. and Raghavendra, C.S., "Minimal full-access networks: enumeration and characterization", *J. Parallel Distrib. Comput.*, **9 (4)**, 1990, pp. 347–356.

[332] Stammermann, A., Kruse, L., Nebel, W., et al., "System level optimization and design space exploration for low power", *in Proc. Int. Symp. System Synthesis*, 2001, pp. 142–146.

[333] Stapl, see http://parasol.tamu.edu/groups/rwergergroup/research/stapl/

[334] STBus Superhighway: Type 3, internal document, ST Microelectronics, 2001.

[335] STBus C++ Class, internal document, ST Microelectronics, 2000.

[336] Stewart, D.B., Schmitz, D.E., and Khosla, R.K., "The Chimera-II real-time operating system for advanced sensor-based robotic applications", *IEEE Trans. Syst., Man and Cybernetics*, **22 (6)**, 1992, pp. 1282–1295.

[337] Stolper, S.A., "Software that travels", *EE Times*, Oct. 1, 2002.

[338] Streamit, see http://cag.csail.mit.edu/streamit/

[339] Stream Processors, RapiDev Tools Suite. Available from http://www.streamprocessors.com/

[340] Stretch, see http://www.stretch.com

[341] Sun, Y., Kumar, S., and Jantsch, A., "Simulation and evaluation for a network on chip architecture using ns-2", *in Proc. Int. IEEE Norchip Conf.*, 2002.

[342] Sutter, H., "The free lunch is over: a fundamental turn towards concurrency in software", *Dr Dobb's Journal*, **30 (3)**, 2005. Available from http://www.gotw.ca/publications/concurrency-ddj.htm

[343] Synopsys Star IP, see http://www.synopsys.com/products/designware/star_ip.html

[344] SUN, see http://www.sun.com

[345] SystemC 2.2, User Manual. Available from http://www.systemc.org

[346] SystemC modeling with the Synopsys Cocentric SystemC Studio, Synopsys, 2002.

[347] SystemC synthesis with the Synopsys Cocentric SystemC Compiler, Synopsys, 2002.

[348] TCC, see http://tcc.stanford.edu

[349] TLM 2.0 specifications. Available from http://www.systemc.org

[350] Tamir, Y. and Frasier, G.L., "Dynamically-allocated multi-queue buffers for VLSI communication switches", *IEEE Trans. Computers*, **41 (6)**, 1996, pp. 725–737.

[351] Tanenbaum, A., *Computer Networks*, Prentice Hall, 1999.

[352] Taylor, M.B., Kim, J., Miller, J., et al., "The Raw microprocessor: a computational fabric for software circuits and general purpose programs", *IEEE Micro*, **22 (2)**, 2002, pp. 25–35.

[353] Taylor, M.B., Lee, W., Miller, J., et al., "Evaluation of the Raw microprocessor: an exposed wire-delay architecture for ILP and streams", *in Proc. IEEE Symp. Comp. Arch.*, 2004.

[354] Tensilica, Inc., "Xtensa instruction set simulator and Xtensa modeling protocol", 2007. Available from http://www.tensilica.com

[355] Thompson, C.D., "Generalized connection networks for parallel processor interconnection", *IEEE Trans. Computers*, **C-27 (2)**, 1978, pp. 1119–1125.

[356] 3DLabs, see http://www.3dlabs.com

[357] Tilera, see http://www.tilera.com

[358] Tiwari, V., Malik, S., Wolfe, A., and Lee, M.T-C., "Instruction level power analysis and optimization of software", *in Proc. Int. Conf. VLSI Design*, **13 (2)**, 1996, pp. 326–328.

[359] Tokuda, H., Nakajima, T., and Rao, P., "Real-time Mach: towards a predictable real-time system", *in Proc. Usenix Machine Workshop*, 1990, pp. 1–10.

[360] Toshiba, see http://www.toshiba.com

[361] Towles, B. and Dally, W.J., "Worst-case traffic for oblivious routing functions", *in Proc. ACM Symp. Parallel Algor. and Arch.*, 2002.

[362] Tullsen, D.M., Eggers, S.J., and Levy, H.M., "Simultaneous multi-threading: maximizing on-chip parallelism", *in Proc. IEEE Symp. Comp. Arch.*, 1995, pp. 392–403.

[363] Urzi, I., Bonner, C., D' Audigier, P., and Sauvage, O., "An HDTV SoC based on a mixed circuit-switched NoC interconnect architecture (STBus/VSTNoC)", *in Proc. IP-based Electr. Syst.*, 2007.

[364] Valiant, L.G. and Brebner, G.J., "Universal schemes for parallel communication", *in Proc. ACM Symp. Theory Comput.*, 1981, pp. 263–277.

[365] Valiant, L.G., "A scheme for fast parallel communication", *SIAM J. Comput.*, **11 (2)**, 1982, pp. 350–361.

[366] Valiant, L.G., "Graph-theoretic properties in computational complexity", *J. Computer Syst. Sci.*, **13 (3)**, 1976, pp. 278–285.

[367] Vangal, S., Howard, J., Ruhl, G., et al., "An 80-Tile 1.28TFLOPS Network-on-Chip in 65nm CMOS", *Int. Solid State Circ. Conf.*, 2007, pp. 98–99.

[368] Veendrick, H.J., "Short-circuit dissipation of static CMOS circuitry and its impact on the design of buffer circuits", *J. Solid-State Circ.*, **SC-19 (4)**, 1984, pp. 468–473.

[369] Venkatraman, V., Laffely, A., Jang, J., et al., "NoCIC: a spice-based interconnect planning tool emphasizing aggressive on-chip interconnect circuit methods", *in Proc. Workshop on System-Level Interconnect Prediction (SLIP)*, 2004, pp. 69–75.

[370] Verhoeff, T., "Delay-insensitive codes: an overview", *Distr. Comput.*, **3 (1)**, 1988, pp. 1–8.

[371] Virtanen, S., Truscan, D., and Lilius, J., "SystemC based object oriented system design", *in Proc. Forum on Design Lang.*, 2001, pp. 1–4.

[372] von Eicken, T., "Active messages: an efficient communication architecture for multiprocessors", *Ph.D. Thesis*, Dept. Electr. Engin. Comp. Sci., University of California, Berkeley, 1993.

[373] VSI Alliance, Virtual Component Interface standard, see http://www.vsi.org

[374] von Eicken, T., Culler, D.E., Goldstein, S.C., and Schauser, K.E., "Active messages: a mechanism for integrated communication and computation", *in Proc. Int. Symp. Comp. Arch.*, 1992, pp. 256–266.

[375] Wang, H.-S., Zhu, X., Peh, L.-S., and Malik, S., "Orion: a power-performance simulator for interconnection networks", *in Proc. Int. Symp. Microarchitecture (MICRO)*, 2002.

[376] Wang, J., Wong, A.K., and Lam, E.Y., "Performance optimization for gridded-layout standard cells", in Staud, W. and Weed, J.T. (eds), *Proc. SPIE, BACUS Symp. Photomask Techn.*, **5567**, 2004, pp. 107–118.

[377] Waksman, A., "A permutation network", *J. ACM*, **15** (**1**), 1968, pp. 159–163.

[378] Weisstein, E.W., "Moore graphs". Available from http://mathworld.wolfram.com/MooreGraph.html

[379] Weste, N. and Eshraghian, K., *Principle of CMOS VLSI Design*, Addison Wesley, 1993.

[380] Wicker, S., *Error Control Systems for Digital Communication and Storage*, Prentice Hall, 1995.

[381] Wiklund, D. and Liu, D., "SoCBUS: switched network on chip for hard real time embedded systems", *in Proc. Int. Parallel and Distr. Proc. Symp.*, 2003, pp. 78–85.

[382] Wiklund, D., "Mesochronous clocking and communication in on-chip networks", *in Proc. Swedish SoC Conf.*, 2003.

[383] Wild, T., Foag, J., Pazos, N., and Brunnbauer, W., "Mapping and scheduling for architecture exploration of networking SoCs", *in Proc. Int. Conf. on VLSI Design*, 2003, pp. 376–381.

[384] Wireless Terminals Solutions Guide, Texas Instruments, 2007.

[385] Wishbone, see http://www.opencores.org/projects.cgi/web/wishbone/wishbone

[386] Whelihan, D., "The NoCSim simulator user guide", Dept. Electr. Comp. Engin., Carnegie Mellon University. Available from http://www.ece.cmu.edu

[387] Wong, C.K. and Coppersmith, D., "A combinatorial problem related to multimodule memory organizations", *J. ACM*, **21** (**3**), 1974, pp. 392–402.

[388] Wu, C.L. and Feng, T.Y., "The reverse-exchange interconnection network", *IEEE Trans. Computers*, **C-29** (**9**), 1980, pp. 801–811.

[389] Xanthos, S., Chatzigeorgiou, A., and Stephanides, G., "Energy estimation with SystemC: a programmer's perspective", *in Proc. WSEAS Int. Conf. Circ.*, 2003.

[390] Xu, J., Wolf, W., Henkel, J., et al., "A case study in networks-on-chip design for embedded video", *in Proc. Design Automation and Test in Europe Conf.*, 1904, pp. 770–775.

[391] Yang, Y. and Masson, G.M., "Nonblocking broadcast switching networks", *IEEE Trans. Computers*, **C-40** (**9**), 1991, pp. 1005–1015.

[392] Yang, P., Wong, C., Marchal, P., et al., "Energy-aware runtime scheduling for embedded multiprocessor SOCs", *IEEE Design and Test of Comp.*, **18** (**5**), 2001, pp. 46–58.

[393] Yu, Z., Meeuwsen, M., Apperson, R., Sattari, O., et al., "An Asynchronous array of simple processors for DSP applications", *in Proc. IEEE Solid State Circ. Conf.*, 2006, pp. 428–429.

[394] Zeferino, C.A., Santo, F.G.M.E., and Susin, A.A., "ParIS: a parameterizable interconnected switch of network-on-chip", *in Proc. IEEE Symp. Integr. Circ. Syst. Design*, 2004, pp. 204–209.

[395] Zeferino, C.A. and Susin, A.A., "SoCIN: a parametric and scalable network-on-chip", *in Proc. IEEE Symp. on Integr. Circ. and Syst. Design*, 2003.

[396] Zhu, X. and Malik, S., "A hierarchical modeling framework for on-chip communication architectures", *in Proc. Int. Conf. CAD (ICCAD)*, 2002, pp. 663–670.

[397] Zivkovic, V.D., van der Wolf, P., Deprettere, E.F., et al., "Design space exploration of streaming multiprocessor srchitectures", *in Proc. IEEE Workshop on Signal Proc. Syst.*, 2002, pp. 228–234.

Index

Printed and bound by CPI Group (UK) Ltd, Croydon, CR0 4YY

28/10/2024

01780086-0001